AN INTRODUCTION
TO THE ELEMENTS OF
AIRPLANE STABILITY
AND CONTROL

AN INTRODUCTION TO THE ELEMENTS OF AIRPLANE STABILITY AND CONTROL

ERIC R. KENDALL, M.SC.

SENIOR TECHNICAL SPECIALIST & ASSOCIATE TECHNICAL FELLOW

THE BOEING COMPANY

BOOK DOMAIN LLC
Publish to Perfection

All inquiries should be addressed to:

Book Domain LLC.
543 E Louise Dr Phoenix, Az 85050

Ordering Information:
Amount Deals. Special rebates are accessible on the amount bought by corporations, associations, and others. For points of interest, contact the distributor at the address above.

Printed in the United States of America.

ISBN-13   Paperback   978-1-964100-09-8
           eBook   978-1-964100-08-1

Library of Congress Control Number: 2024919210

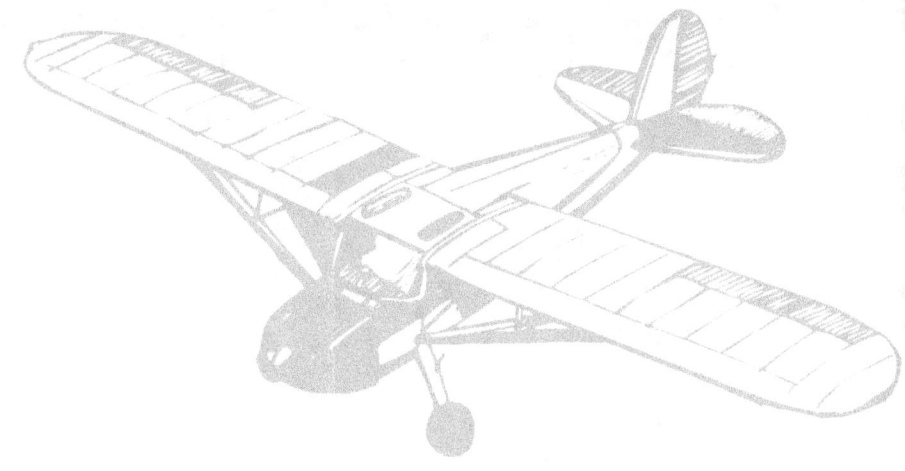

The basic elements of airplane stability and control are presented as a text in plain words with minimal use of mathematics. The purpose is to attract high-school seniors and college freshmen, who have an interest in airplanes, to a subject that they may wish to study at a more advanced level as a degree choice.

When provided with suitable data, the student will be able to attempt meaningful work on airplane configuration and have a clear understanding of the technical issues involved.

Exposure to a text of this more elementary level will assist student and teacher in progressing through the more advanced levels to be taught in a full degree course.

Laymen, private pilots and aero-modelers may also benefit from the material presented.

ABOUT THE AUTHOR

ERIC R. KENDALL began his career in 1949 in the U.K. as an apprentice in the airplane industry. He studied Aeronautical Engineering at Southampton University and received a scholarship from the Society of British Aircraft Constructors for post graduate studies at the College of Aeronautics, Cranfield. He earned a M.Sc. degree majoring in Aerodynamics and was awarded the Woods of Colchester Prize for his academic performance.

At the Air Registration Board, he studied human mortality statistics and proposed the 10^{-7} per flight hour commercial airplane safety standard commonly accepted to-day. At Smiths Aviation Division, he worked on the first commercial automatic landing system for the De Havilland Trident airplane.

He emigrated in 1966 to work with the Lockheed Georgia Company in Flight Dynamics on the C-5A Program. Then, in 1981 he joined the Learjet Corporation as Chief of Aerodynamics and Propulsion where he participated in the design of external airframe modifications to improve handling qualities of the Model 25 and 35 airplanes. On the Model 25, "sharks teeth" attached to the wing leading edge prevented severe wing drop at stall. On the Model 35 highly swept ventral fins prevented locked-in deep stall and improved dutch roll damping. He reviewed the Piaggio GP-180, "Avanti", airplane for potential U.S. marketing and received the

AIAA Best Paper Award for his lecture to the Wichita Section on "The Aerodynamics of Three-Surface Airplanes".

In 1987, Eric joined the McDonnell Douglas Company in Long Beach, California as Senior Manager of Flying Qualities on the C-17A "fly-by-wire" airplane. He also began a 20-year period of teaching "Airplane Stability and Control", "Aerospace Propulsion", "Airplane Design" and "Avionics" giving evening lectures to graduate and under-graduate students at the California State University at Long Beach.

He completed his career in 2011 as a Senior Technical Specialist and Associate Technical Fellow of the Boeing Aircraft Corporation in Long Beach, California.

CONTENTS

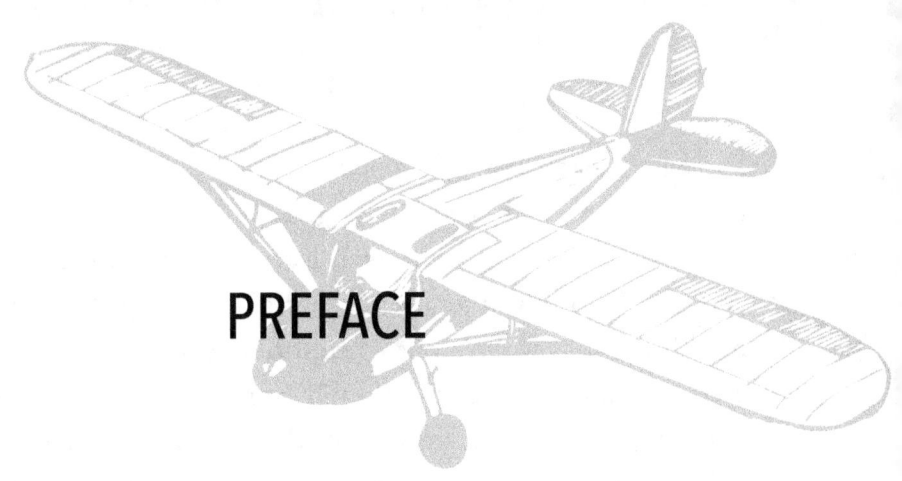

PREFACE

WHY DO PEOPLE GO TO AIR SHOWS?

A hundred years ago people went to air shows to witness the miracle of a man in flight. Nowadays, airplanes are taken very much for granted. They have been around since before we were born, but air show attendance is still high. Some 535,000 people attended Oshkosh in 2010 to see 10,000 airplanes. Some 36,000 attendees camped out. Interest in the latest and fastest airplanes draw the crowds; In these crowds of uplifted faces, several "gonna-be" pilots or "wanna-be" airplane designers are eager to get out of high-school and start their careers.

WHY READ ABOUT AIRPLANE STABILITY AND CONTROL?

Well, for the "gonna be" and "wanna be" readers it will help to prepare you for one of the courses you will study in college. Perhaps also you may get more enjoyment when you go to airshows because you will be able to see details that go un-noticed to the untrained eye. Just like birds who have kept their secret for millions of years, airplanes do not show all onlookers the attributes that enable them to fly smoothly and maneuver so gracefully.

In the early days before man learned how to fly, he could get enough lift to leave the ground, but after leaving the ground, the machine would wobble out of control and immediately return to earth causing damage, injury and sometimes death to its pilot. Balancing and steering was identified as a principal challenge to our ability to fly. The subject of stability and control teaches us about balancing and steering.

Why must I wait to go to college before I can learn more about this "invisible quality?

Well, to move through college at a reasonable pace and to become a professional pilot or aeronautical engineer, you need to understand a level of applied mathematics that is higher than that taught in high-school.

But a large part of the stability and control subject can be taught using clear explanations with a small amount of high school math. In fact, your formal study of college textbooks on the subject will benefit from first reading a simpler text like this one that is less dependent on mathematics.

So, if you are a young person who wishes to make a career in the aerospace industry, either as a pilot, an aeronautical engineer or a research scientist, this book will help you get started on a path towards understanding how airplanes are designed to be stable and controllable in flight.

It may be read also by the hobbyist or by the layman who simply has an interest in airplanes and wants to get more pleasure from attendance at air shows.

Eric Kendall
1st February 2011

CHAPTER 1

Equilibrium and Stability

1.1 Simple Systems

Before exploring the stability of airplanes it is helpful to review the stability of some simpler devices. We might first consider a pendulum hanging on a pivot as shown in figure 1(a) below:

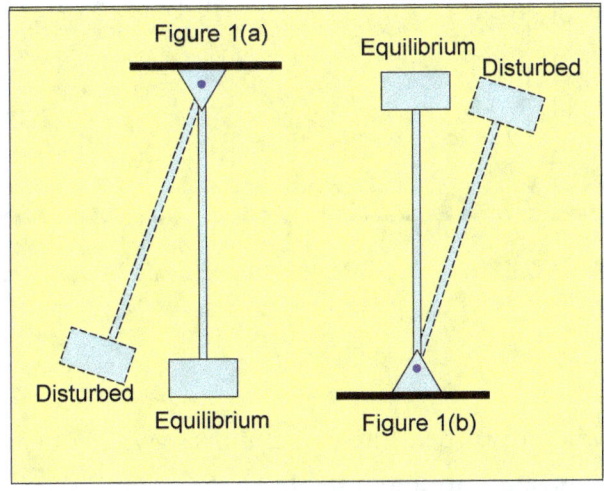

Figure 1.1

When hanging on a rigid rod straight down below its pivot, it is in equilibrium. It will stay in equilibrium so long as it is not disturbed. When it is disturbed as shown and released to move freely, it will return to equilibrium. This pendulum is stable with respect to the equilibrium position shown.

When situated straight up on a rigid rod above its pivot as shown in figure 1(b), it is in equilibrium. It will stay in equilibrium so long as it is not disturbed. When it is disturbed as shown and released to move freely, it will move away from equilibrium and has no tendency to return. This pendulum is unstable with respect to the equilibrium position shown. It will continue to move away from this equilibrium position until it finds a new one like the pencil shown in figure 1.2.

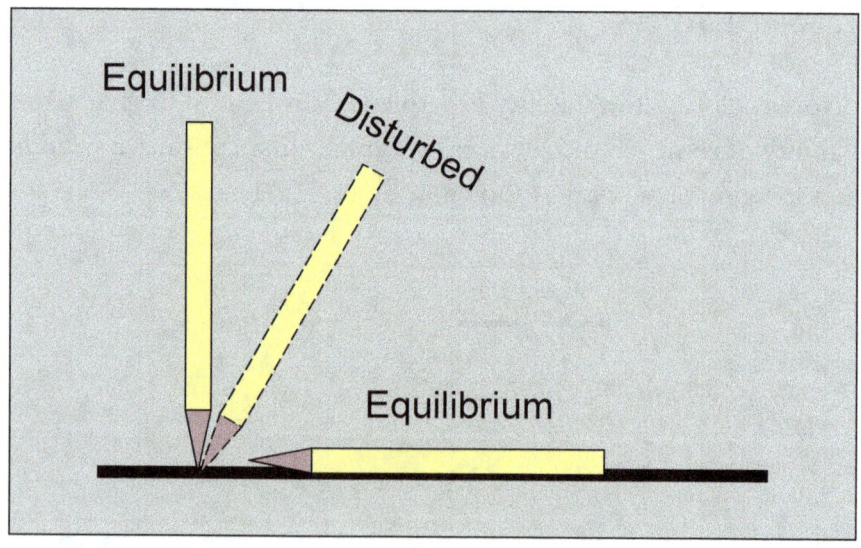

Figure 1.2

Here, when disturbed, the pencil moves away from an unstable equilibrium position to find a stable one.

We note that any discussion of the stability of a system must be referenced to an equilibrium condition. So, as part of learning about the stability of an airplane, we must first learn how to place it in equilibrium.

The next chapter shows how to do this after explaining why the early experimenters experienced so much difficulty.

CHAPTER 2

THE LONGITUDINAL STABILITY OF A WING

2.1 GLIDING ON A WING.

The reader is asked to imagine a wing descending through the air so that, in side view, it appears as shown in the figure below:

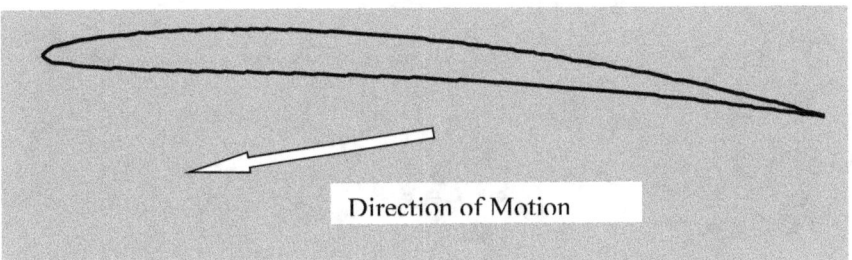

Direction of Motion

Figure 2.1

The wing's cross section is a smooth slender shape tapering towards its trailing edge, inclined leading-edge-up with respect to its direction of motion, and curved slightly so that a mean line between its upper and lower surfaces curves above a straight line drawn between its leading and trailing edges. This slight upward curvature is referred to as positive camber.

2.2 CAMBER AND INSTABILITY.

Positively cambered wing sections were found by Otto Lilienthal to provide higher lift and lower drag than un-curved, (flat), sections. He built and flew gliders in the 1890's and considered that, in these cambered sections, "lay the whole secret of the art of flying".

Yet, during his gliding flights, he found that the center of air pressure on the wing moved fore and aft as his glider pitched nose-up and nose-down so that he was forced to constantly move his body weight to control and stabilize his glider. A photograph taken during one of his gliding flights shows his struggles to keep his weight aligned by raising his legs as the center of lift moved forward on the wing. Quite a strenuous exercise! Current-day "Kite-Skaters" will be able to understand the effort needed to steer in strong gusty wind conditions.

Figure 2.2 Otto Lilienthal Gliding in 1895
(Ref: Google, "Otto Lilienthal Glider Pictures")

A few years later in 1901, Wilbur and Orville Wright, who performed manned glider experiments before making the first powered flight in 1903, also found that positively cambered wing sections were very unstable. The brothers accepted this as a challenge that pilot practice would have to overcome. In their opinion, pilots would have to be trained to fly unstable airplanes just as cyclists learned by practice to balance on a bicycle. They owned a bicycle business in Dayton.

Dr. Samuel Langley, who was also researching in this field, virtually abandoned hope that there would ever be a solution to this "obstacle to manned flight" caused by the de-stabilizing fore and aft movement of the center of lift on a wing.

2.3 Movement of the Center of Lift.

In his 1901 lecture to the Society of Western Engineers, (Ref.1), Wilbur Wright discussed the issue at some length:

> *"The balancing of a gliding or flying machine is very simple in theory. It consists in causing the center of gravity to coincide with the center of pressure. But in actual practice there seems to be an almost boundless incompatibility of temper which prevents their remaining peaceably together for a single instant, so that the operator, who in this case acts as a peacemaker, often suffers injury to himself while attempting to bring them together. If a wind strikes a vertical plane, the pressure on that part above the center will exactly balance that on the other side. But if the plane be slightly inclined, the pressure on the part nearest the wind is increased and the pressure on the other part decreased, so the center of pressure is now located, not in the center*

of the surface, but a little toward the side which is in advance. If the plane be still further inclined the center of pressure will move still further forward,..."

The center of pressure movement being described corresponded to predictions from Joessel's formula for square flat plates, and the inclination being referred to in the lecture is the inclination away from the vertical. The formula can be represented by the graph shown in Figure 2.3 below where x_p/c is the distance of the center of pressure back from the leading edge of the flat plate divided by the chord length of the plate and α is the angle between the airflow direction and the chord, referred to as the "angle of attack". We see that when the plate is held normal to the airflow direction, ($\alpha = 90$ deg.) the center of pressure is in the middle. ($Xcp/c = 0.5$). Then, as the plate is tilted with its upper edge moving forward, the angle between the plate and the airflow direction is reduced and the center of pressure moves forward. When the plate is almost parallel to the airflow direction at an angle of attack almost equal to zero, Joessel's formula predicts that the center of pressure will be close to 20% of the chord length behind the leading edge.

The direction of movement agreed with trends found experimentally by Avanzini in 1804 and had been used for predicting wind loads on flat sloping roofs.

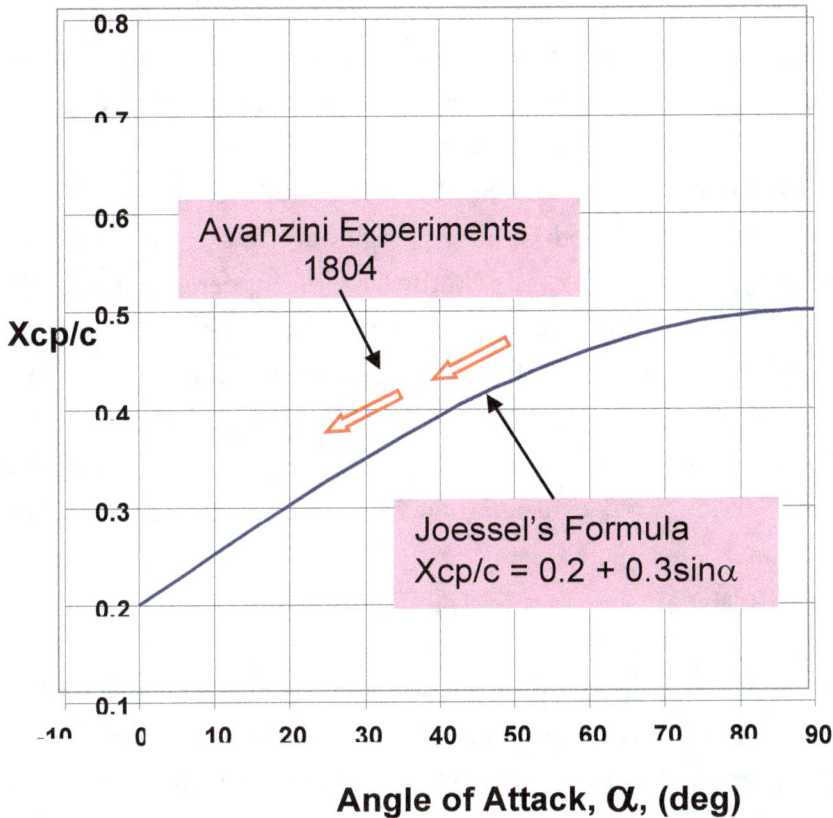

Figure 2.3

There appeared to be some belief that this flat plate formula would apply also to the slightly cambered, (almost flat), shapes of the chord-wise sections being used on the early gliders. But it does not support the existence of instability for wings gliding at low angles of attack. If a glider flying at an angle of attack of 5 degrees is pitched up to 6 degrees the lift would increase and, according to the formula, would move aft from about 22.5% to 23% of the wing chord. This would give rise to a nose-down moment to oppose the pitch-up disturbance and provide a stabilizing influence.

So, if these early glider pilots were working hard against an instability that could not be explained by Joessel's formula, then apparently, the formula did not apply to cambered surfaces in the low angle of attack region in which the gliders were flying.

When comparing the flow past flat and cambered surfaces, Lilienthal had noted that, at low angles of attack, the air passing an inclined **flat** plate was disturbed on the upper surface after it passed over the sharp leading edge but it flowed smoothly over the **curved** surface of a positively cambered airfoil, following its contour completely. The curved-down leading edge of the wing met the oncoming air more tangentially so that the positively cambered airfoil section was considered to be more "conformal" to the airflow.

2.4 THE WRIGHT BROTHERS EXPLORE FURTHER.

After encountering significant nose-down pitch and diving tendencies on their 1901 glider designed with more camber than that on the glider they had flown in 1900, the Wright brothers concluded that *"the trouble was due to a reversal of the direction of travel of the center of pressure at small angles."* As an experiment to confirm this, they flew the upper wing of their bi-plane glider as a kite and noted that, at low angles of attack, the center of air pressure moved **aft** as angle of attack was **reduced**. Wilbur's account of the experiment is copied below:

> *"We had removed the upper surface from the machine and were flying it in a wind to see at what angles it would be supported in winds of different strengths. We noticed that in light winds it flew in the upper position shown in the figure, with a strong upward pull on the cord c. As the wind became stronger the angle of incidence became*

less, and the surface flew in the position shown in the middle of the figure, with a slight horizontal pull; but when the wind became still stronger it took the lower position shown in the figure, with a strong downward pull. It at once occurred to me that here was the answer to our problem, for it is evident that in the first case the center of pressure was in front of the center of gravity, and thus pushed up the front edge; in the second case they were in coincidence and the surface in equilibrium, while the third case the center of pressure had reached a point even behind the center of gravity, and there was therefore a downward pull on cord.

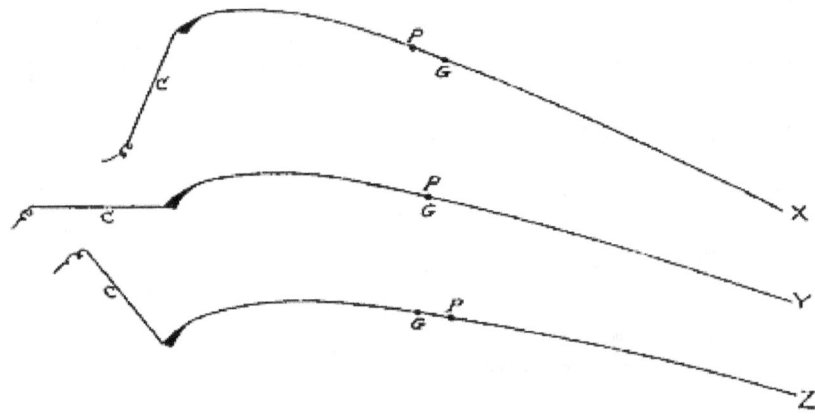

P, center of pressure; G, center of gravity.
(From Ref 1)

The experiment had shown that, to support the weight of the wing as the wind speed increased, the angle of attack decreased and the center of pressure moved aft. So, in this range of low angles of attack, the Joessel 'flat plate' formula did not apply to cambered surfaces.

By reducing the camber on their 1901 glider, the Wrights were able to control its pitching motion more easily and went on towards their 1903 historical powered first flight.

Then, in 1905, Wilbur Wright's diary recorded his wind-tunnel measurements of the center of pressure movement on cambered surfaces. The results are shown in the next figure in comparison with those prescribed by Joessel's formula.

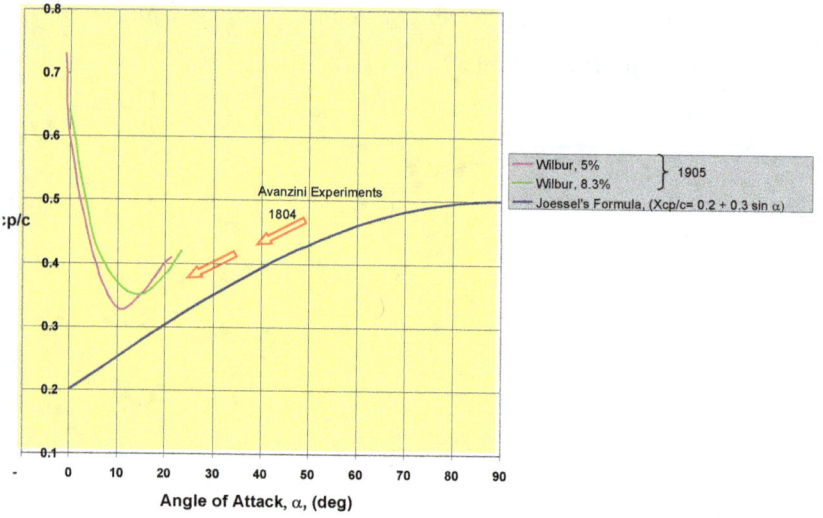

Figure 2.4

This indicates that the center of pressure movement on a cambered airfoil does not begin to follow Avanzini's trend and Joessel's flat plate formula until an angle of attack between about 10 and 15 degrees is exceeded. At higher angles the airflow will no longer follow the curved airfoil contour so the flow pattern will be more similar to the flat-plate flow.

Wilbur's data can now be used to explain the instability being experienced.

2.5 FLYING ON A POSITIVELY CAMBERED WING.

Using the low angle of attack information for positively cambered airfoils, we can perform a simple analysis to show why the early glider experimenters had difficulty in maintaining balance, after they became airborne with their weight initially aligned below the center of pressure.

We start by plotting a graph of x_p /c vs. angle of attack for a particular airfoil, say Wilbur's 5% cambered airfoil, and assume that the airplane's center of gravity is located at some point on it.

In Figure 2.5 below, the airplane's center of gravity is located 45% of the chord length back from the wing leading edge where it is **in balance** with the location of the center of pressure at an angle of attack of about 4 degrees. This 'in balance' condition is more easily seen in Figure 2.6.

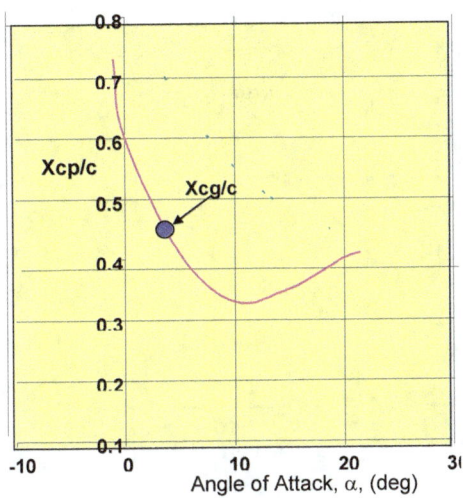

Figure 2.5

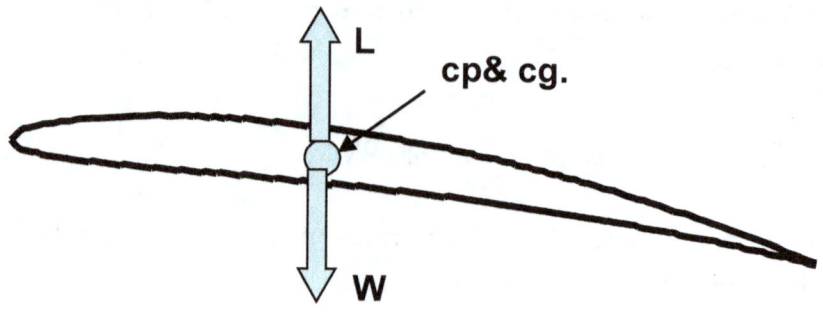

Figure 2.6

Now, if the angle of attack is disturbed from 4 to 5 degrees, then the lift will increase and the center of pressure will move forward to 40% of the chord with the center of gravity remaining at 45% of the chord as shown in Figures 2.7. The wing is now unbalanced as shown in Figure 2.8.

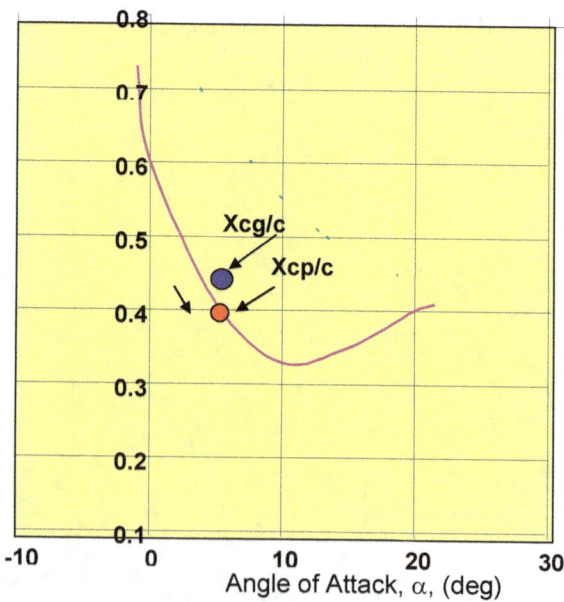

Figure 2.7

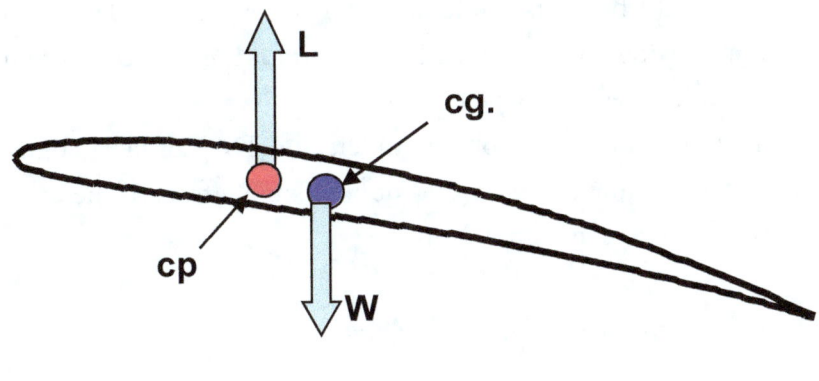

Figure 2.8

Unless the pilot responds quickly with a nose-down corrective moment the wing will continue to pitch up and reach the angle of attack at which the wing stalls.

We can see that, when flying a positively cambered airfoil, a condition of balance where the center of gravity is initially aligned with the center of pressure at any angle of attack below the stall angle will become unbalanced following any small disturbance in angle of attack. Like the pencil balanced on its pointed tip, the positively cambered airfoil cannot be placed in **stable** equilibrium. It will require continued pilot inputs to maintain balance in flight.

From this simple study we can more readily visualize the basic reason for Otto Lilienthal's difficulty and perhaps for his fatal accident in 1896 when he was apparently unable to move his weight forward fast enough or far enough to suppress a strong nose-up disturbance probably caused by a large gust of wind.

Also we can understand clearly why Wilbur Wright, in his lecture to The Western Society of Engineers in 1901, likened his gliding experience with trying to ride a wild horse.

The Wright brothers implemented a controllable surface ahead of the main wing and practiced hard to attain the skill to maintain balance under these unstable conditions.

Finally we might also better understand Dr. Samuel Langley's frustration with the unsteady and destabilizing effects of the center of pressure movements on his wing.

2.6. THE AERODYNAMIC FORCE ON A WING

By studying more closely the variation of lift and center of pressure over a range of angles of attack used in flight, we can discover a characteristic of airfoils that removes entirely the need to keep track of the center of pressure movement in our stability studies.

In Figure 2.9 below, the chord-line AB that joins the centers of curvature of the leading edge, A, and the trailing edge, B, of a section of a wing is inclined at an angle of attack, α, to its direction of motion through the air.

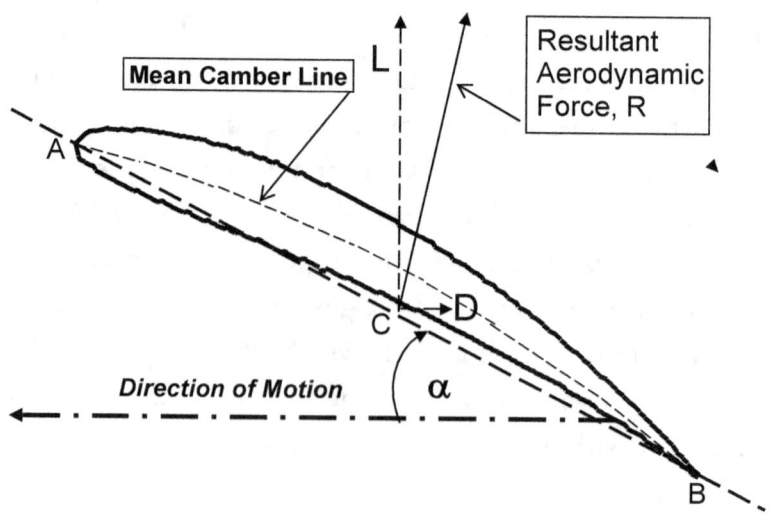

Figure 2.9

The aerodynamic pressures acting on the wing are equivalent to a single resultant force, R that acts through point C on the chord The resultant force, R, can be resolved into a *lift* force, **L**, that acts normal to the direction of motion, and a *drag* force, **D**, that acts parallel to and in a direction opposite to the direction of motion.

The wing section shape has *positive camber* noted from the fact that the mean camber line, situated midway between its upper and lower surfaces, is curved upward *above* the straight-line chord AB.

A measure of the camber is the maximum distance between the mean camber line and the chord line expressed as a percentage of the chord length.

The variation of lift, L, and center of pressure location with angle of attack are shown in Figure 2.10 together with airflow visualization photographs taken from Ref. 2.

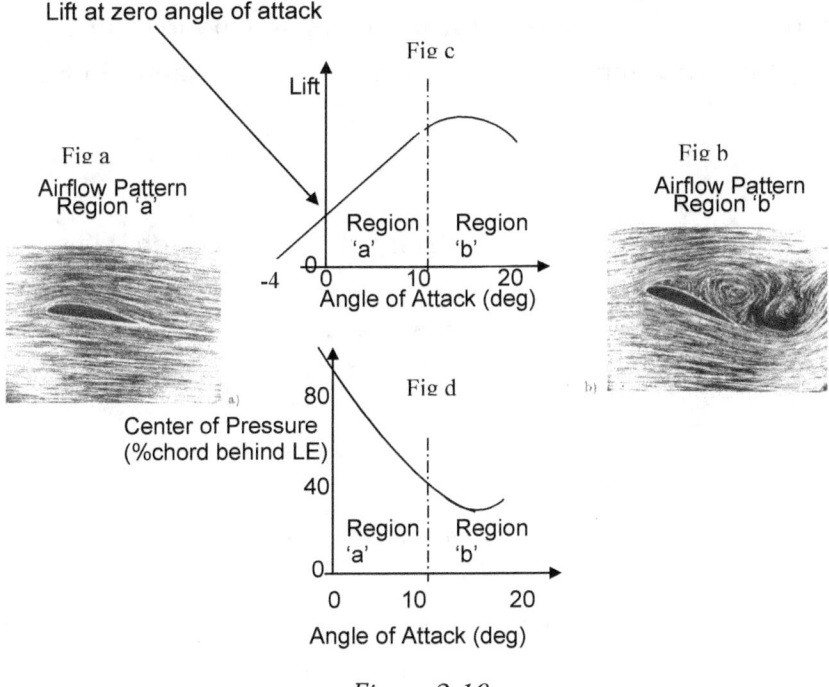

Figure 2.10

2.7 Lift Variation

At low angles of attack, in region 'a', air flows smoothly past the airfoil as shown in Fig a on the left, and lift increases linearly with angle of attack as shown in Fig c. Then, as angle of attack is increased further into region 'b', the air flow is unable to follow the airfoil's upper surface and begins to separate as shown in Fig b on the right. In region 'b' the lift reaches a maximum and begins to decline as shown in Fig c. The airfoil has 'stalled'.

2.8 "Zero-Lift" Condition

The positively cambered airfoil will produce positive lift at zero angle of attack as shown in Fig c and must be set to a negative angle of attack for zero lift. Here, "zero lift" is comprised of negative lift on the forward section with equal and opposite positive lift on the aft section resulting in a pure nose-down couple as shown in Figure 2.11.

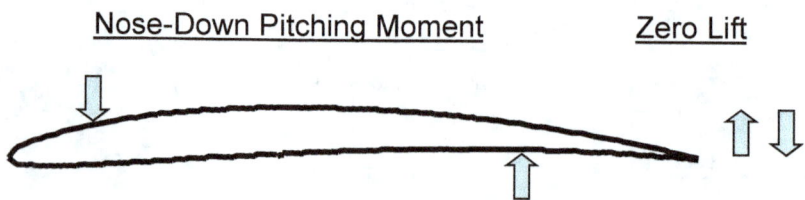

Figure 2.11

2.9 Center of Pressure Location.

In the low angle of attack region, the center of pressure on a positively cambered airfoil is located aft, as shown in Fig 2.10d, and moves forward on the airfoil as angle of attack is increased. Then, in the higher angle of attack region where the airfoil begins to stall, the forward motion of the center of pressure begins to reverse having almost reached the 30% chord location.

2.10 Effect on Stability

These data were not available to the early experimenters of manned gliders. Otto Lilienthal was experimenting during the late 1800's, and the Wright brothers were attempting to fly in the early 1900's.

The airflow visualization pictures in Figure 2.10 were first published by L. Prandtl in 1910, seven years after the Wright's first successful powered flight in 1903. Wilbur Wright first reported the center of pressure chord-wise movement from cambered airfoil tests in his wind tunnel in 1905.

2.11 The Aerodynamic Center.

We have seen that as angle of attack is increased, the lift on a wing moves forward and grows larger. This is depicted in Figure 2.12.

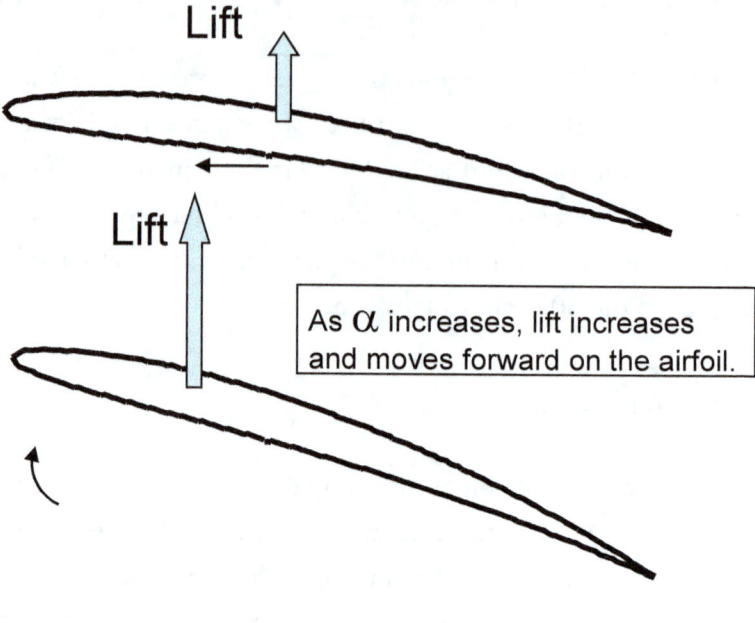

Figure 2.12

Now, in order to discuss the moment change on the airfoil caused by this lift movement, we must identify the point on the air-foil about which the pitching moment is being assessed. The pitch-ing moment, M, depends on the moment reference point chosen.

2.11a) *Moment about the Trailing Edge.* As angle of attack increases, the lift force and its distance from the trailing edge **both increase**. So the pitching moment about the trailing edge will experience an increase in the nose-up direction due to increases in **both** the lift and its moment arm as shown by the line '1-2' in Figure 2.13.

Point '1' corresponds to the angle of attack for zero lift where, as seen in figure 2.11, the pitching moment for a positively cambered airfoil has a negative value.

2.11b) *Moment about the Leading Edge.* The moment about this reference point will be in the nose-down direction as shown by line '1-3' in the figure. It starts at the same point '1' since, at zero lift, the pitching moment must be a pure couple and a pure couple can be placed anywhere on the airfoil. In this case, the lift force increases but the moment arm decreases. So the line '1-3' is not as steep as the line '1-2'.

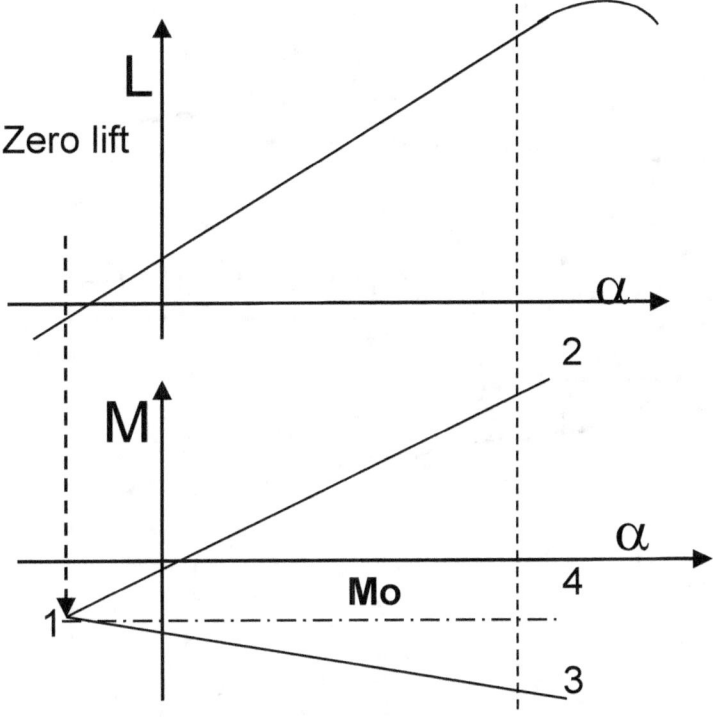

Figure 2.13

On the figure we can draw a line '1-4' having zero slope between lines '1-2' and '1-3'. Line '1-4' then corresponds to a point somewhere on the airfoil between its leading and trailing edge where the moment has a value that does not change as α changes.

This point is called the aerodynamic center of the airfoil.

As can be seen from figure 2.13, the moment about this important reference point is equal to the moment at zero lift, **Mo**.

Relative to this point, the product of lift, L, (that grows as α increases) and distance, d, behind the point, (that reduces as α increases), remains constant. This is depicted in Figure 2.14.

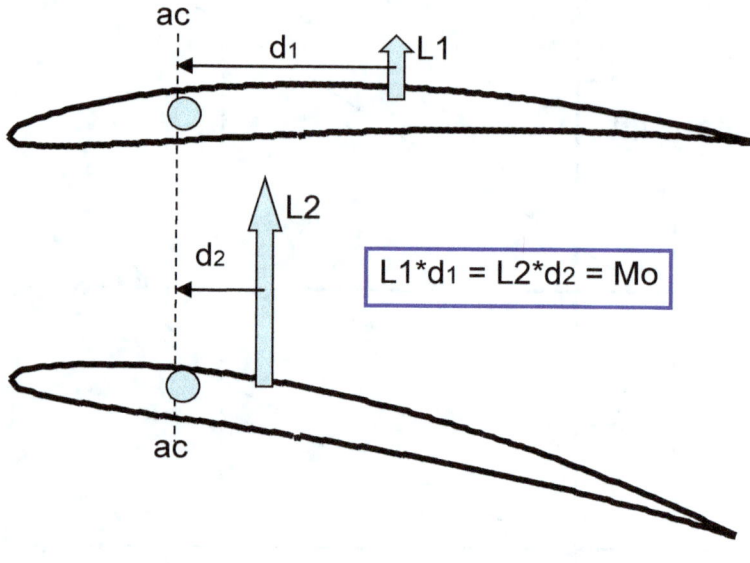

Figure 2.14

*(Thin Airfoil Theory provides a mathematical proof that the aerodynamic center is located at **one-quarter chord** distance back from the leading edge. The result is confirmed by wind-tunnel tests performed on thin airfoils.)*

Because of this finding we no longer need to consider the movement of the center of pressure in our stability studies. We can replace the 'moving' lift by a lift force situated at the aerodynamic center that varies in magnitude as angle of attack changes and by a constant moment, Mo, acting about the aerodynamic center.

As angle of attack changes, the moment will remain constant and the lift will change only in magnitude.

This arrangement is shown in Figure 2.15

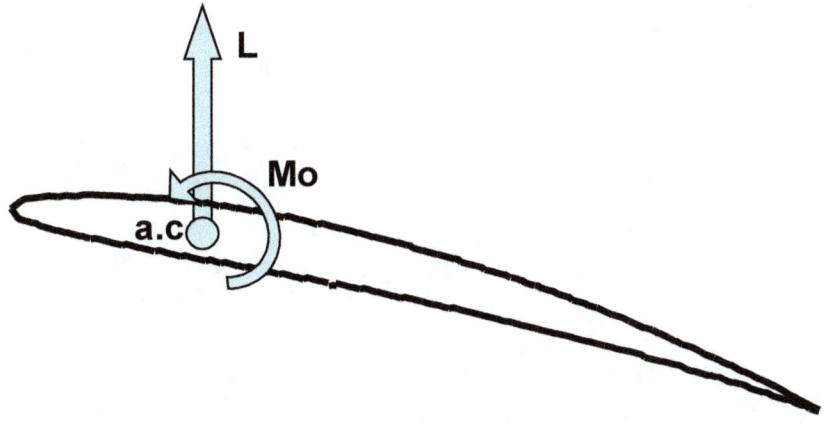

Figure 2.15

Now it is easier to assess the equilibrium and stability of an airfoil and to proceed towards assessing the equilibrium and longitudinal stability of a complete airplane.

CHAPTER 3

ADDING THE TAIL

3.1 LIFT, BALANCE AND STABILITY.

In 1893 Albert Zahm noted that, **if the lift acted ahead of the weight**, any disturbance that caused the angle of attack to change would cause the lift to change in a direction that would make the disturbance grow. Thus, an angle of attack increase would cause a lift increase which, acting ahead of the weight, would give rise to a nose-up pitching moment about the center of gravity. This would cause the angle of attack to increase further. The reverse would be true if the initial disturbance was a reduction in angle of attack. In this case, a lift decrease, acting ahead of the weight, would cause a reduced nose-up moment about the center of gravity causing the angle of attack to decrease further.

The only solution to this pitching instability would be to ensure that **the weight is always placed ahead of the lift.** With the center of gravity ahead of the aerodynamic center the lift change following an angle of attack disturbance would tend to reduce the disturbance.

But now we have a problem if we want to fly on positively cambered airfoils because the zero-lift moment, Mo and the lift-weight couple both act in the nose-down direction as seen on the left of figure 3.1. The airfoil alone on a straight untwisted wing cannot be placed in equilibrium.

Albert Zahm showed that an incremental downward force acting further back on the wing/airplane as shown on the right of figure 3.1 would be needed to maintain equilibrium in the undisturbed state.

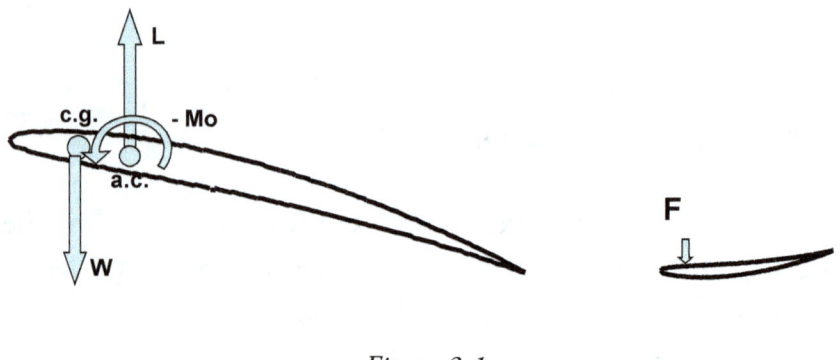

Figure 3.1

A small down-force on a tail set well back from the wing produces a balanced condition. Then we have:

Lift + Down-force = Weight   L + F = W
Moment about center of gravity is zeroMcg = 0

Also, a disturbance that causes an increase in angle of attack will produce a restorative nose-down pitching moment due to the increased wing lift aided by a change of tail lift in the upward direction.

The airplane resists angle-of-attack change, $\Delta M / \Delta \alpha < 0$

Conventional airplanes are designed, as shown in the figure, with a smaller aft tail surface to maintain balance. The tail is set at a leading-edge-down incidence angle relative to the wing and produces a small down force on a long moment arm to balance the nose-down moments being produced by the main wing.

Note: *This is a direct implementation of Albert Zahm's 1893 requirement for stable equilibrium in flight.*

3.2 DECALAGE.

The difference between the zero-lift incidence of the wing and the zero-lift incidence of the tail is called the **decalage angle** and is positive when the forward wing incidence is leading-edge-up with respect to the tail incidence,

On the simplest of toy gliders close examination of the slots in the fuselage for holding the wing and the tail will reveal a positive decalage angle as shown in Figure 3.2.

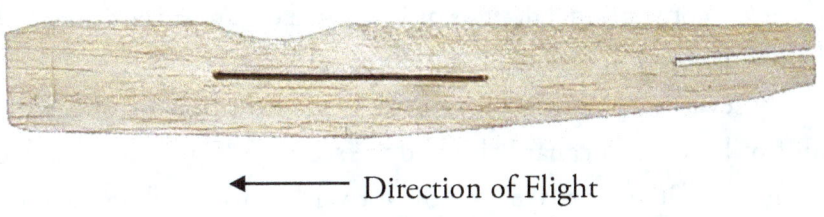

← Direction of Flight

Figure 3.2

This model will be stable in flight to the left so long as a small weight is attached to its nose to place the center of gravity ahead of the wing's quarter chord location.

The completed model is shown in figure 3.3 below:

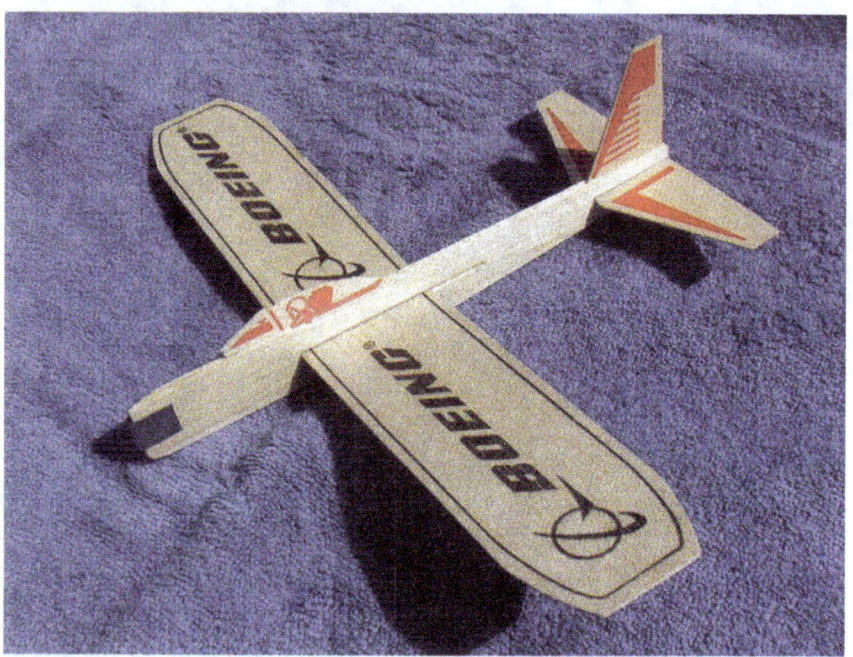

Figure 3.3

3.3 Canard Designs

Some less conventional designs place a small canard surface ahead of the main wing to produce an 'up' force for balance.

By reversing the direction of the fuselage of our toy glider the tail slot becomes a canard slot and we see from Figure 3.4 that we still have positive decalage with the forward surface leading-edge-up relative to the aft surface.

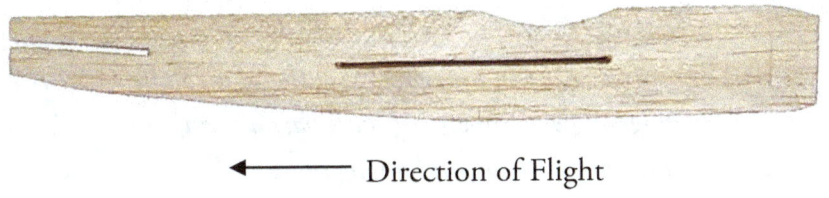

◄─────── Direction of Flight

Figure 3.4

In this configuration lift on the forward canard surface will increase when angle of attack increases causing an incremental destabilizing effect. So the center of gravity must be placed well ahead of the wing's aerodynamic center to maintain inherent stability.

The completed model is shown in figure 3.5 below:

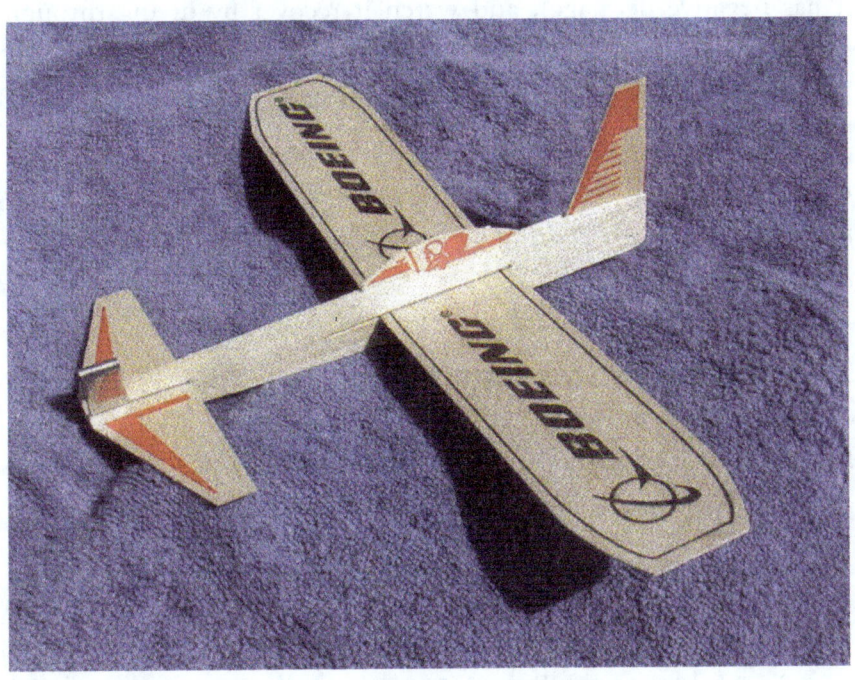

Figure 3.5

Notes:

1) *The student may purchase a model kit for a couple of dollars and can benefit in an understanding of airplane stability by intelligently experimenting with various configurations.*

2) *Credit for the model and photographs is due to Joe Kelly, student of "Airplane Stability and Control", CSULB, Spring, 2005*

It is interesting that the small decalage angle is barely noticeable on the assembled model yet, like Zahm's requirement, which seemed to go un-noticed by the early manned-glider researchers, it is essential for the successful flight of any airplane.

As proof, launch the toy glider inverted, (in which condition it has negative decalage), and watch it recover by diving through 180 degrees to level flight in the other direction where it reaches its upright balanced condition with positive decalage.

Note: The inverted launch should be high enough to prevent the glider from striking the ground during it recovery to its "positive decalage" condition.

3.4 THE REQUIREMENT FOR POSITIVE Mo.

In the conventional and canard glider configurations described above we have satisfied three basic requirements for successful flight.

1. Positive lift, $L > 0$
2. Equilibrium, $M = 0$ and..
3. Inherent stability, (*Nose-down pitching moment when lift increases*)

When these three requirements are combined as shown on Figure 3.6, it can be seen that they infer an important associated requirement.

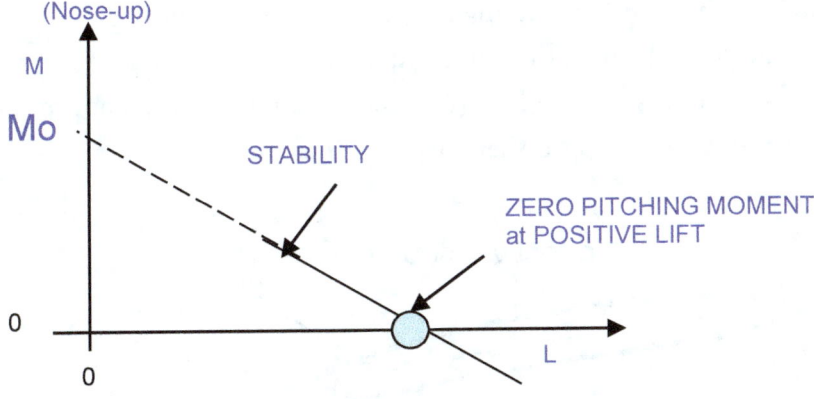

Figure 3.6

By extrapolating the requirement for stability back to the zero-lift condition, we see that the zero-lift pitching moment, Mo, for the **complete airplane** must be positive.

Mo > 0

Positive decalage produces this condition for the complete airplane even though it uses a positively cambered wing for high lift and low drag. The combination of a wing and a tail, or a canard and a wing, set at incidences to provide a positive decalage, will blend with the contour of a negatively cambered surface for which the aft part is curved upward.

Thus Zahm's requirement and the requirement for a positive value of Mo are one and the same. They are satisfied by the implementation of positive decalage.

3.5 TRAILING EDGE FLAPS

Mo of the wing will have a large **negative** value when its positive camber is increased considerably by lowering trailing edge flaps to obtain high lift at low landing speeds as shown in Figure 3.7.

The down-load on the tail required to balance the airplane under these conditions will be large and the tail must be sized appropriately and set at an appropriate negative incidence.

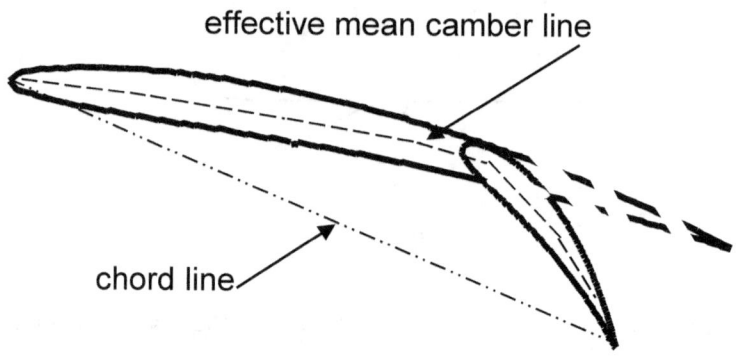

effective mean camber line

chord line

Figure 3.7

3.6 THE NEUTRAL POINT

A second surface, added behind the wing, will generate a change of lift, ΔL_T when angle of attack is increased. Additional lift may be generated also from the fuselage, engine nacelles, etc. It is convenient to lump these "additional" components with the wing lift increase to produce a "**wing/body**" lift increase, ΔL_{WB}, acting at the wing-body aerodynamic center.

When the angle of attack of the whole vehicle is increased by an amount, $\Delta\alpha$, the wing-body lift will increase by an amount ΔL_{WB}, and the tail lift will increase by an amount ΔL_T.

The total lift increase is:

$$\Delta L_{ToT} = \Delta L_{WB} + \Delta L_T$$

Now there will be a point between the aerodynamic center of the wing-body and the aerodynamic center of the tail at which the total change of lift causes no change in pitching moment. The nose-up moment about this point due to the increase in wing-body lift exactly balances the nose-down moment due to the increase in tail lift. This location is referred to as the "neutral point", (**n.p.**), of the complete airplane. It is a key reference point for determining where the total airplane center of gravity must be located to establish a stable equilibrium condition.

When the airplane weight is located ahead of the neutral point and the tail lift is adjusted to balance the airplane in pitch, then it is in stable equilibrium just like the pendulum in figure 1.1(a).

However, when the weight is located behind the neutral point and balanced with the appropriate tail lift, it is in unstable equilibrium just like the pendulum in figure 1.1(b).

The neutral point may be determined from data sheets or from measurements made on a model in a wind tunnel.

3.7 SUMMARY

Figure 3.8 below illustrates the stable equilibrium arrangement of forces and moments acting on an airplane of conventional wing/tail configuration:

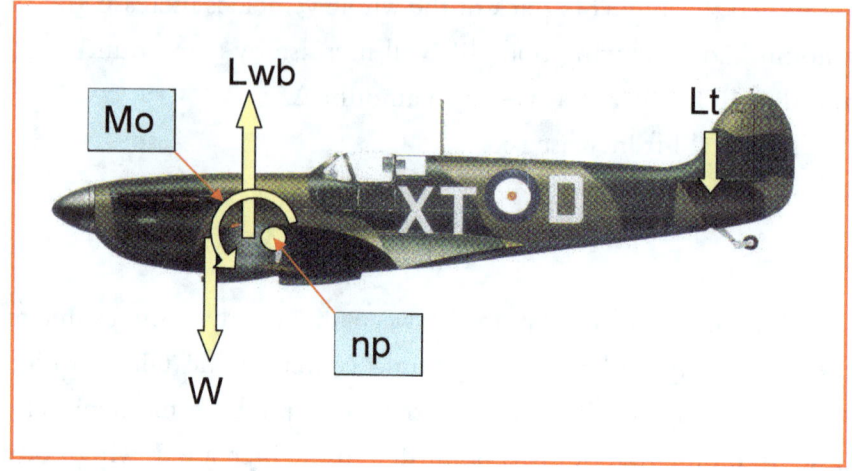

Figure 3.8

- Wing-Body lift, **LwB**, acts through the wing-body aerodynamic center.
- A zero-lift moment, **Mo**, acts around the wing-body aerodynamic center. *(It will normally be negative i.e. directed nose-down)*
- Tail lift, **LT**, acts at the tail's aerodynamic center. *(It will normally be negative i.e. directed downward)*
- The neutral point, **n.p.** is between the aerodynamic centers of the wing-body and the tail.
- Weight, **W**, acts through the center of gravity ahead of the neutral point, **n.p.**
- For pitch stability, the weight, W, may be behind the wing-body aerodynamic center but must be ahead of the airplane's neutral point.

CHAPTER 4

LONGITUDINAL STATIC STABILITY

4.1 CENTER OF GRAVITY LOCATION

Now that we know how to balance the airplane as shown in figure 3.8, we can study its stability by disturbing it. Suppose we disturb its angle of attack by an increased amount Dα. Then, if it is flying in region 'a' of the lift versus α curve shown in figure 2.10 the lift will increase. And, if the airplane's weight is ahead of the neutral point the additional lift will resist the angle of attack increase and tend to restore the airplane to equilibrium. But if the weight acts behind the neutral point, the increased lift will cause the airplane to continue to pitch up away from its equilibrium angle of attack. Clearly, with its center of gravity ahead of the neutral point the airplane is in stable equilibrium just like the pendulum in figure 1.1(a). And if the airplane's center of gravity is behind the neutral point, then the airplane is in unstable equilibrium just like the pendulum in figure 1.1(b). If the angle of attack is increased by an isolated upward vertical gust, then the stable airplane will return to its original angle of attack while immersed in the gust or once the gust encounter has passed.

This static stability disappears, or becomes "neutral" when the center of gravity is located **at** the "neutral point". With the center of gravity located at the neutral point, lift change due to angle of attack change acts through the center of gravity and does not produce a moment change about it.

4.2 The Static Margin

The amount of static stability possessed by an airplane increases as the center of gravity is moved further forward. A measure of stability may then be taken as the distance of the center of gravity ahead of the neutral point so long as we adjust the measurement by a reference distance that accounts for the size of the airplane. So, we measure the distances of the neutral point, Xnp and the center of gravity, Xcg back from the wing leading edge, as shown in figure 4.1, and divide each measurement by the wing chord length, c.

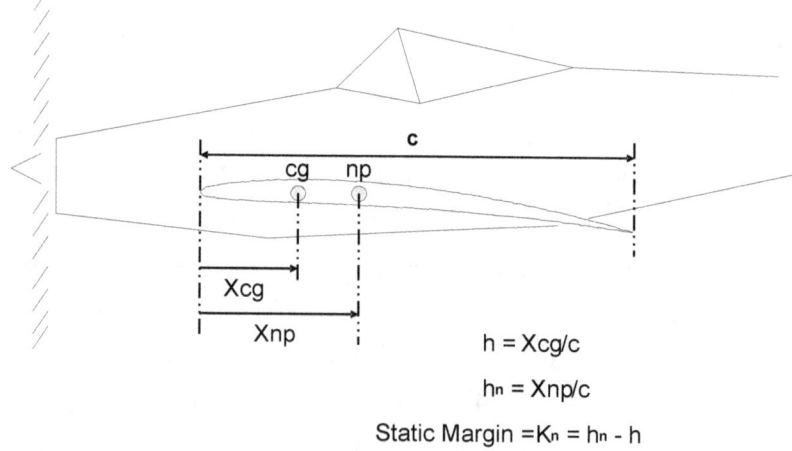

$$h = Xcg/c$$

$$h_n = Xnp/c$$

$$\text{Static Margin} = K_n = h_n - h$$

Figure 4.1

Then we have hn = Xnp/c, h = Xcg/c so that the measure of static stability is hn – h. This measure is called the Static Margin, Kn.

$$\text{Static Margin} = \text{Kn} = \text{hn} - \text{h}$$

The static margin measures how well the airplane resists angle of attack changes. The larger the static margin the stronger the resistance.

But if the resistance to angle of attack change is too high it will be difficult to maneuver the airplane in pitch. Then the pilot will not like its controllability characteristics. Clearly, too much stability is to be avoided.

4.3 STATIC STABILITY GRAPHS

The static longitudinal stability of an airplane can be assessed from a graph of pitching moment, M versus angle of attack, α or lift, L.

With the airplane center of gravity ahead of the airplane's neutral point the increased lift, ΔL due to an angle of attack increase will cause a negative incremental pitching moment, ΔM and the quantity ΔM/ΔL will be negative for positive static stability. The distance between the airplane's neutral point and its center of gravity increases as the center of gravity moves forward. So, the negative value of ΔM/ΔL will increase indicating a higher level of static stability. This quantity can be seen as the slope of a graph of M versus L as shown in figure 4.2 below:

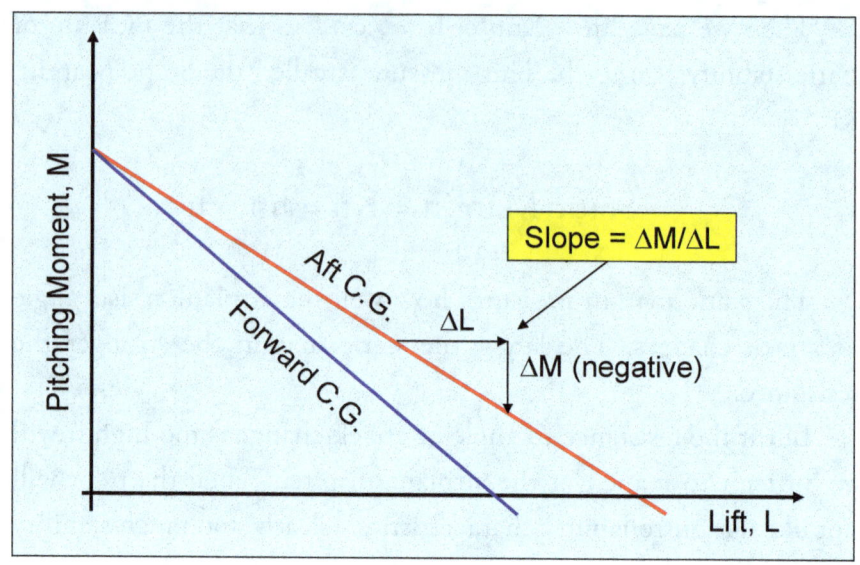

Figure 4.2

CHAPTER 5

Longitudinal Maneuverability

5.1 The Effect of Pitch Rate

The resistance to maneuvering comes from the static stability where lift acting at the neutral point increases as angle of attack increases **and** from the fact that the airplane is flying around a curved path as shown in the figure below:

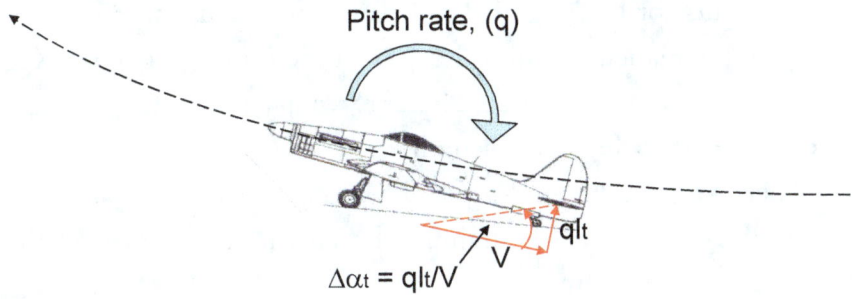

Figure 5.1

In order to fly this curved path the airplane must continually pitch up at a rate of 'q' radians per second so that the tail, that is situated a distance 'lt' feet behind the center of gravity, is moving downward at a velocity of 'q*lt' feet per second relative to the airplane's c.g. As shown by figure 5.1 this increases the tail angle of attack by an amount 'Dat' over the value corresponding to steady level flight where the pitch rate is zero. The increased tail angle of attack due to pitch rate produces an incremental increase of tail lift that adds resistance to the maneuver. This extra resistance is equivalent to an aft shift of the neutral point when maneuvering in the pitch axis. The further aft point is then called the airplane's maneuver point. When the airplane's center of gravity is situated at this point, no control inputs are needed to produce or to sustain a pitching maneuver. This could be very dangerous causing the airplane to maneuver so violently that structural failure occurs.

5.2 THE MANEUVER MARGIN

A measure of the airplane's resistance to maneuvers may then be taken as the distance of the center of gravity ahead of the maneuver point so long as we adjust the measurement by a reference distance that accounts for the size of the airplane. So, we measure the distances of the maneuver point, Xmp, and the center of gravity, Xcg behind the leading edge of the airplane's wing and divide each measurement by the wing chord length, c.

Then we have hm = Xmp/c, h = Xcg/c so that the measure of resistance to maneuvers is hm − h. This measure is called the Maneuver Margin, Hm

$$\text{Maneuver Margin} = Hm = hm - h$$

Though more is to be learned about airplane maneuverability, it is necessary first to learn a little about the airplane's primary controls that are used by the pilot to perform maneuvers.

We need to ensure that the control force required to initiate a maneuver can be comfortably generated by the pilot and that the force does not become unacceptable as the maneuver progresses.

CHAPTER 6

The Primary Controls

6.1 Three-Axis Control System

The primary controls on an airplane are used by the pilot to control the pitching, rolling and yawing motion of the airplane.

During their gliding experiments at Kitty Hawk the Wright Brothers developed a canard surface for pitch control, differential outer-wing twist for roll control and an aft-mounted vertical surface for yaw control. They were the first to develop a three-axis control system for an airplane.

A photograph of their 1901 glider is shown in the next figure.

Figure 6.1
(Ref: Google, "Wright Brothers Glider Pictures")

As airplane development progressed, pitch control was implemented using a hinged surface called an **elevator** attached to an aft tail. Roll control used hinged surfaces called **ailerons** attached to the outer part of the wing, and yaw control was provided by a hinged surface called a **rudder** attached to the aft part of a vertical stabilizer.

These surfaces, the ailerons, elevator and rudder, are shown on the Westland Wyvern airplane in the next figure.

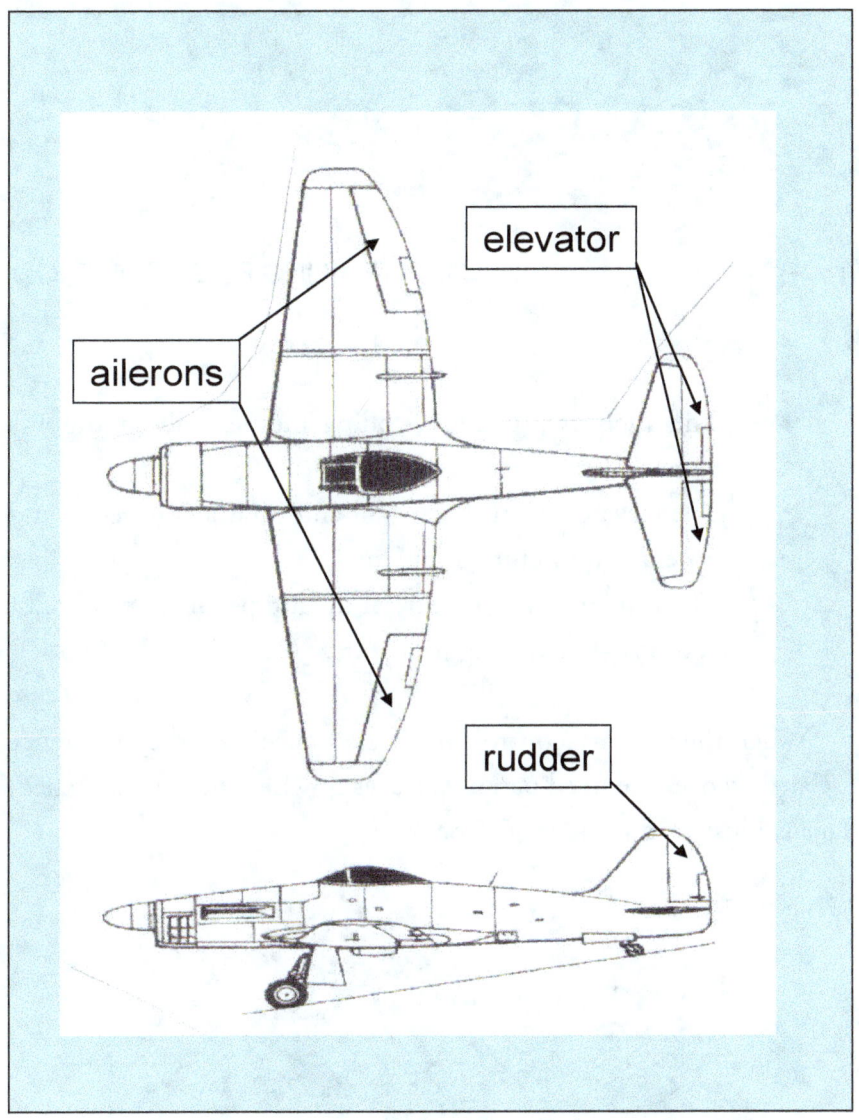

Figure 6.2

In their neutral position the control surfaces form part of the fixed surface's cross sectional profile as shown below:

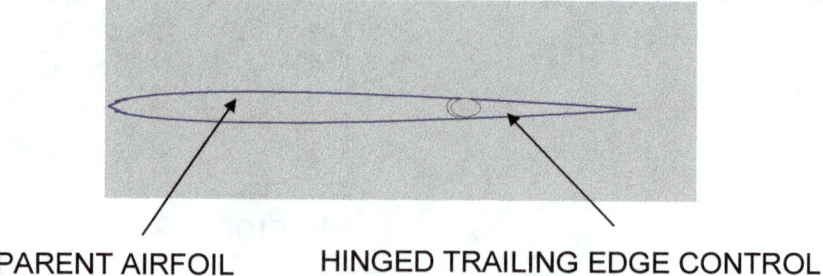

PARENT AIRFOIL   HINGED TRAILING EDGE CONTROL

Figure 6.3

- The ailerons produce a rolling moment about the airplane's centerline.
- The elevator is on the aft tail and produces nose-up and nose down pitching moment.
- The rudder is on the vertical fin and produces nose-right and nose-left yawing moments.

When the primary control surfaces are displaced they influence the airflow over the fixed parent surfaces to which they are attached. This is illustrated in figure 6.4 below.

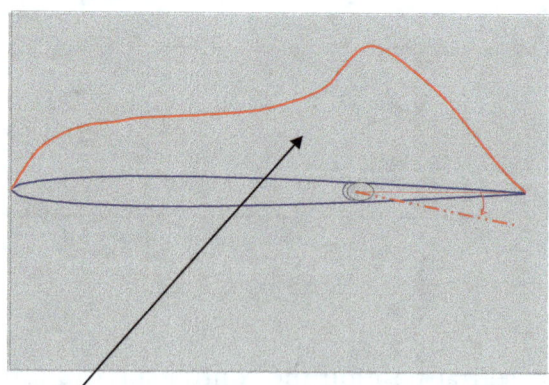

Pressure change due to control deflection

Figure 6.4

6.2 Pilot Effort Required

In airplanes where mechanical linkages connect the pilot's control column to the control surface as shown simply in figure 6.5 pilot's force alone is used to rotate the control surface about its hinge. Here the control column is shown connected to the elevator control surface.

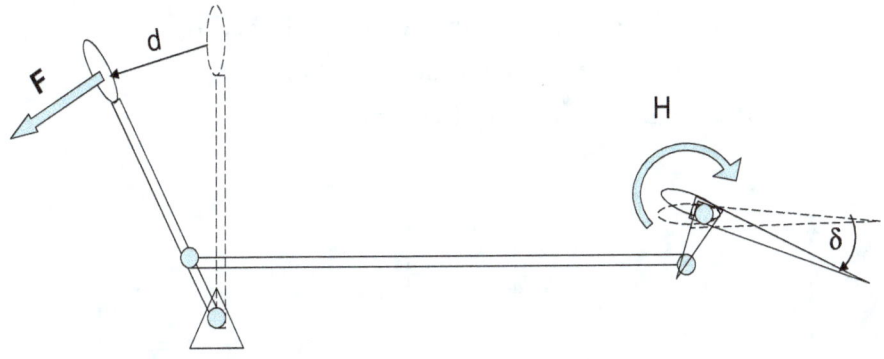

Figure 6.5

The pilot's force, **F** and the applied hinge moment, **H** will increase as the stick deflection, **d** and the control surface rotation, δ both increase. The work done by the pilot in moving the control stick through a distance, d will be equal to that done in rotating the control surface through an angle, δ by applying a hinge moment, H against the aerodynamic resistance.

The work done by the pilot is equal to the area under the graph of 'F' versus 'd' and, assuming that the force, F increases linearly with the stick displacement, d the graph is as shown on the left of figure 6.6 below. And if the hinge moment, H increases linearly with the surface rotation, δ, then its graph is as shown on the right in figure 6.6. Then if negligible work is done by the pilot against control system friction, the areas of both triangles are the same and:

Work done is(1/2)*F*d = (1/2)*H*d

Where:

F = pilot's force, (lb)
d = column deflection at point of applied force, (ft)
H = control surface hinge moment, (ft-lb.)
δ = control surface rotation, (radians)

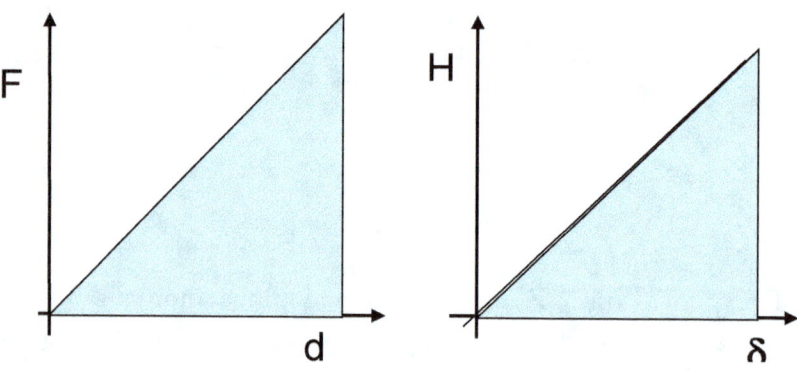

Figure 6.6

The maximum column travel is limited by the pilot's arm length and is about sixteen inches from the back stop to the front stop of his control column. The maximum control surface rotation from full trailing edge up to full trailing edge down is limited to about forty degrees. Larger rotation angles would cause airflow separation on the surface leading to a loss of aerodynamic effectiveness.

When these stick displacement and surface rotation limits are inserted into the previous equation for work done, it yields the following relationship between pilot force, F and control surface hinge moment, H for a pitch control system:

$$F*16/12 = H*40/57.3$$

Or

$$F = 0.52*H$$

This relationship varies only slightly between airplanes in which the pilot's manipulator is a control column or stick. From it we can identify an important issue to be confronted in the design of airplane primary control systems.

6.3 THE EFFECT OF SIZE AND SPEED.

The force, F that can be applied comfortably by the pilot and the maximum force that can be applied and held for a short time have been determined and are specified for various types of airplane in design requirements.

But the hinge moment, H increases as airplane size and speed increase and a point is reached beyond which even the most ingenious control surface designs will result in unmanageable pilot forces.

An approximate effect of airplane size and speed on the aerodynamic hinge moments of the control surfaces can be illustrated by comparing the product of wing area, S wing chord, c and airspeed squared for a range of airplanes of varying size and speed as shown in the following table.

Table 6.1

Airplane	Cessna	Learjet	MD-80	DC-10	B-747
S (sq.ft)	174	232	1209	3647	5500
c (ft)	4.9	7.0	13.2	24.6	27.3
V (ft/sec)	219	558	575	625	673
ScV^2	41M	510M	5300M	35,000M	68,000M
Ratio	1	12	129	854	1658

M = Million,

Ratio = $ScV^2/41M$ and shows the approximate size of aerodynamic hinge moments on a particular airplane compared to the Cessna value.

We note from the last line of the above table that the effect of airplane size and speed on control surface hinge moments is considerable. But the designer of each airplane must find a way to ensure that the pilot control forces required to maneuver the airplane will be acceptable.

6.4 PILOT FORCE REQUIREMENTS

To proceed further into the problem, we can select one of the several force requirements that appear in the design specifications for military and commercial airplanes of different categories and derive an approximate quantitative measure for the magnitude of hinge moment that corresponds to the force requirement.

Let's consider a "stick force per 'g' requirement" that states that the pilot force 'F' must not exceed 120 lb. to initiate or maintain a maneuver along a curved path withan incremental normal acceleration equal to '1g' above that for level flight. This is to avoid pilot fatigue during maneuvers.

From page 61. F = 0.52*H So the magnitude of H must not exceed 120/0.52 = 231 ft.lb.

Now if:

Air pressure acting on the control surface is $(1/2)pV^2$ lb./sq.ft.
Control surface area is S sq.ft.
Control surface chord is C ft.
Then the hinge moment can be written as:

$$H = Ch * (1/2)\, pV^2 Sc$$

Where Ch is a non-dimensional coefficient that depends on the shape, hinge location and deflection, δ of the control surface and on the angle of attack, α of the parent surface, (tail, wing or fin), to which it is attached.

So we can write it as:

$$Ch = Ch_\delta * \delta + Ch_\alpha * \alpha$$

Where $Ch\ \delta$ = the amount the hinge moment coefficient changes per unit of control deflection and $Ch\ \alpha$ = the amount the hinge moment coefficient changes per unit of angle of attack change of the parent surface.

At the beginning of a maneuver, before α changes from its equilibrium (trimmed) flight value, the hinge moment is due only to Ch_δ

Then

$$H = H_{(\delta)} = Ch_\delta * (1/2)\, pV^2 Sc * \delta$$

The aerodynamic hinge moment, $H_{(\delta)}$ must resist the control displacement δ otherwise the surface would not remain centered when the pilot's control is released. Instead, it would be rapidly driven to its limit by the aerodynamic hinge moment acting on it. To avoid pilot fatigue, $H_{(\delta)}$ must be no more negative than -231 ft.lb. and,

$$Ch_\delta \geq \frac{-231}{(1/2)\, pV^2 Sc * \delta}$$

Estimated values for the larger airplanes in the previous table are tabulated below:

Table 6.2

Airplane	Learjet	MD-80	DC-10	B-747
$Ch_{\delta\,LIMIT}$	-0.502	-0.036	-0.003	-0.0016

These values for the DC-10 and the B-747 are so small that small changes in control surface contour due to production inaccuracies, damage, ice accretion or even bird droppings might cause Ch_δ to become positive causing control surface divergence and airplane structural failure. To prevent such an event, a control surface design limit is:

$$Ch_\delta \leq -0.10 \, / \, radian$$

So on the DC-10 and B-747, control surfaces that conformed to this limit would produce pilot control forces approximately 30 to 60 times greater than the specification requirement for "stick force per 'g'".

Hydraulic power is needed to move the control surfaces on these large fast airplanes. Their control systems are briefly described in Chapter 8.

We note that the Learjet easily conforms to the limit. Its control surfaces are mechanically connected to the pilot's control column.

But there are many airplanes larger than Learjets and smaller than DC-10s that can avoid the use of hydraulically-powered controls by correct and sometimes ingenious selection of control configurations. The MD-80 is one such airplane that we will visit later.

This takes us to the threshold of a long history and towards a large volume of information on how to design control systems so that the pilot forces required to move the control surfaces without the use of hydraulic power remain well within his/her strength capability.

The next chapter presents a brief review of this subject.

CHAPTER 7

Un-Powered Primary Controls

7.1 Resistance to Control Movement

In the early 1940s during World War II, airplane development and production were progressing rapidly. Larger airplanes were in demand to carry larger bomb loads and to fly faster to and from the target. As airplane size and speed increased, the hinge moment, H became larger and required much attention to the design of control surfaces that would encounter less resistance to rotation about their hinges. At that time it was considered that hydraulic power systems were not sufficiently reliable to be trusted with the control of airplanes. Failure would almost certainly lead to significant control difficulties and probable loss of the airplane.

However, it was anticipated then, that airplane speed and size would continue to increase and that reliable hydraulic systems would eventually be developed for airplanes that could be conceived to exist in the more distant future. Until then designing manually activated control systems was a challenge to be confronted for the larger, faster airplanes of the day.

The principal problems and the various ingenious solutions are well described in the aeronautical literature. References 3 and 4 are recommended for further reading.

The following sections of this chapter will describe the basic issues to be considered and present some approximate guidelines for the design of control surfaces in manually activated control systems.

7.2 Stick Forces and Pilot Feel

If the pilot pulls back on the stick to increase angle of attack, the force required to move the elevator trailing edge upward may change during the maneuver. We remember that $Ch_\alpha * \alpha$ forms part of hinge moment coefficient, Ch. If the elevator tends to float trailing edge up due to the changing aerodynamic pressures on it that arise from the angle of attack increase, (i.e. if Ch_α is negative), the pilot will feel that less stick force is required to sustain the maneuver than was needed to initiate it. The controls will lighten up as the maneuver develops. If this effect is excessive, it will be assessed as objectionable by the pilot like a car with severe over-steer tendencies.

Appropriate adjustment of the control surface geometry can reverse the float direction and cause stick force to gradually increase as the maneuver develops. This would be much preferred so long as the force increase is not excessive, causing an objectionable under-steer condition.

We note that choosing the correct control surface geometry and control system arrangement is important to the airplane's control feel just as are the chassis adjustments that are made to "INDY" cars during a race. We note too, that choice of control geometry is important to produce the desired amount and direction of control surface float when angle of attack changes. We need to choose a control configuration that will provide the desired value of Ch_α

Desired force characteristics are prescribed in design specifications for acceptable handling qualities of various types of airplanes and much data exists on Ch_δ and Ch_α for various control geometries.

7.3 THE EFFECT OF CONTROLS ON STABILITY.

An elevator control surface forms the aft portion of the tail as shown in figure 7.1 below and is connected to the pilot's pitch control stick. If the pilot's stick is not held fixed when an angle of attack disturbance occurs, the freely hinged elevator surface will float to a new equilibrium angle and affect the airplane's resistance to the angle of attack change. If it floats trailing edge up when angle of attack increases, it will cause a reduction in the resistance to the angle of attack change and the static margin will be less than the margin with the surface fixed.

Clearly, on airplanes where the elevator control surface can float about its hinge due to an angle of attack change, we must distinguish between a "stick fixed" and "stick free" static margin.

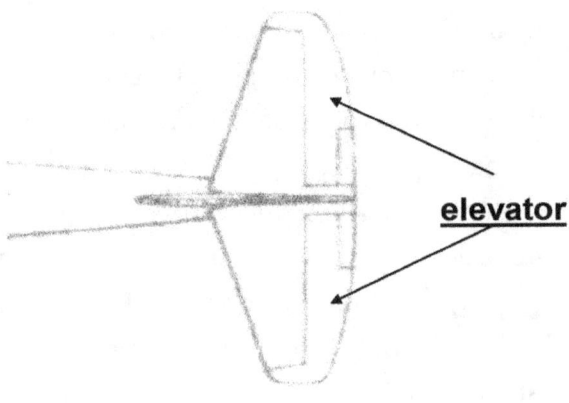

Figure 7.1

So   Stick-Fixed Static Margin = Kn = hn − h

And   Stick-Free Static Margin = Kn' = hn' − h

The stick free neutral point, hn' <hn if the elevator floats trailing edge up when angle of attack is increased.

7.3.1 STICK FIXED

If the pitch control surfaces of the airplane remain fixed when the airplane is held away from its "in-balance" angle of attack by an amount Dα, then the stick fixed static stability can also be assessed from a measure of the incremental pitching moment change, ΔM that resists the angle of attack displacement, Dα. The slope of the graph will then provide a measure of the "stick fixed" static longitudinal stability.

7.3.2 STICK FREE

Now suppose that the control column is free so that the elevator surface can rotate freely on its hinge. The elevator will rotate to a new angle to find equilibrium about its hinge under the changed aerodynamic moment acting about the hinge. The amount and direction of this free surface rotation will depend on the surface geometry and the pressures acting on it.

If the elevator floats trailing edge up when angle of attack is increased, it will produce an incremental nose up pitching moment that reduces the airplane's resistance to the incremental angle of attack being held. So, ΔM/ΔL will have a lower negative value and the stick free static stability will be less than the stick fixed static stability.

When an increased level of stick free static stability is desired, surface geometry can be altered to provide more surface area ahead of the hinge. This changes the surface's **aerodynamic balance** and can be sufficient to cause the elevator to float trailing edge down when angle of attack is increased. This increases the airplane's resistance to an angle of attack change and provides a level of stick free stability that is better than the level of stick fixed stability.

These effects on the M versus L graphs are shown in Figure 7.2 below:

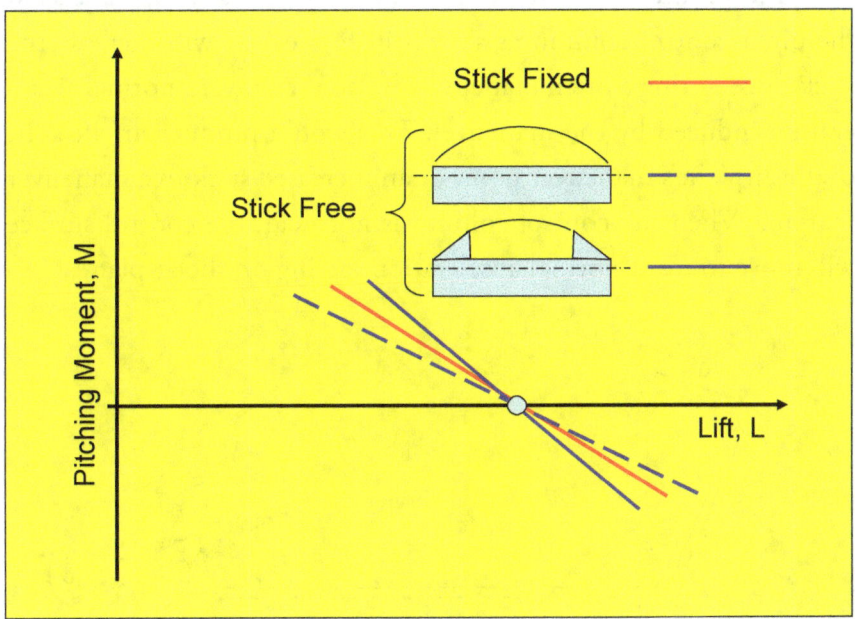

Figure 7.2

Note: The legend shows elevator surface geometries corresponding to two different stick free graphs. The lower surface has more area ahead of the hinge line and has the potential to increase the stick-free static stability above the stick-fixed level.

7.4 EFFECT OF CONTROL SYSTEM MASS BALANCE

When the control column is not held and the pitch control surfaces are able to float during an angle of attack change, the maneuverability will be modified by the free control surface in a manner similar to that explained for the stick-free static stability margin.

Also, since maneuvering introduces accelerations on the airplane, any mass imbalance of the control system will introduce forces that tend to move the control system. These forces can be created by design to produce a desired effect. A bob-weight attached ahead of the pitch control column as shown in figure 7.3, will increase the stick force during a pitch maneuver due to the upward normal acceleration induced by the maneuver. This would produce an aft shift of the airplane's maneuver point or an increased stick free maneuver margin. When the control column is not held, the control surface will move to resist normal accelerations acting on the airplane.

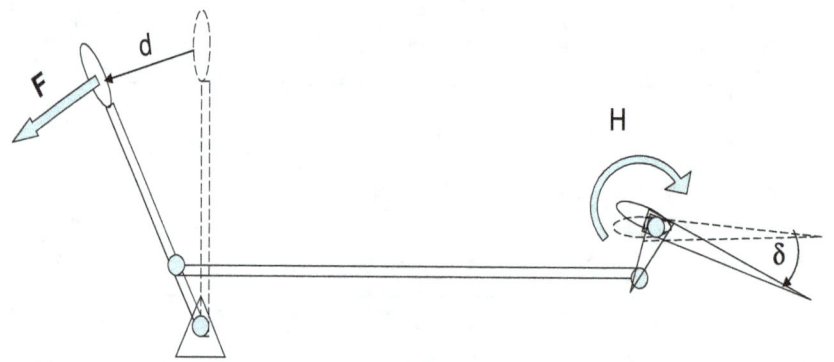

Figure 7.3

Whether the stick-free control inputs are influenced by elevator float due to angle of attack change or to maneuver accelerations, in either case, a stick free maneuver point, hm' can be established for the airplane and its control system design.

Then the stick free maneuver margin will be:

$$\text{Stick Free Maneuver Margin} = Hm' = hm' - h$$

7.5 CONTROL SURFACE DESIGN

Several ways of controlling the amount and direction of control surface float, (i.e. the size and sign of Ch_α), have been developed and are well proven. They are well reported in the literature.

Here we will consider only two of them; the horn balance and the beveled trailing edge. Their effectiveness in solving a severe control problem will be demonstrated.

7.5.1 HORN BALANCE.

A horn balance adds area ahead of the hinge at the tip(s) of the control surface. When angle of attack is increased with the stick-free, the powerful suction on the horn balance will cause an elevator to float trailing edge down.

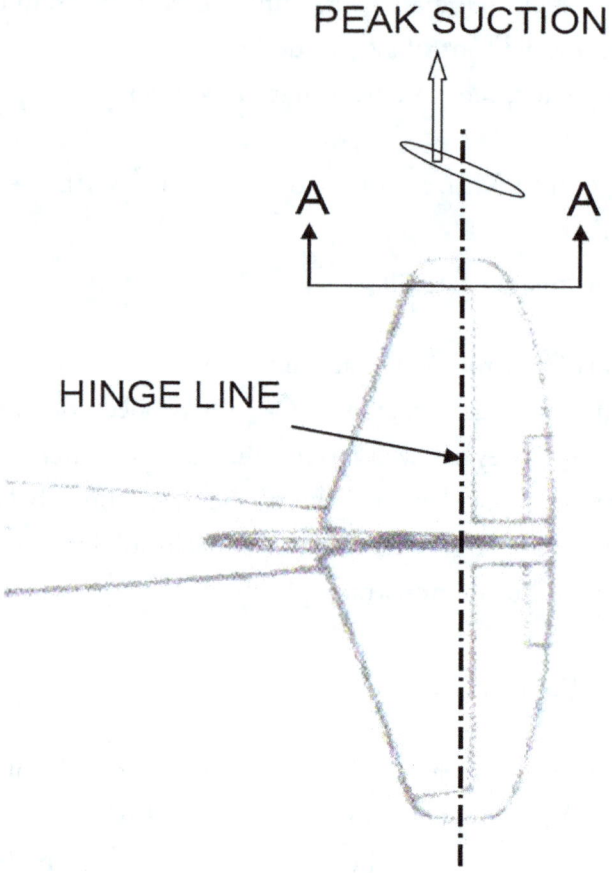

Figure 7.4

The horn balance is very effective in making Ch_α more positive without large reductions in the magnitude of Ch_δ

7.5.2 BEVELED TRAILING EDGE

Shaping the trailing edge of the control surface to have a less acute angle,(see figure 7.5), has an effect on Ch_δ and Ch_α similar to that produced by the horn balance. Both become less negative.

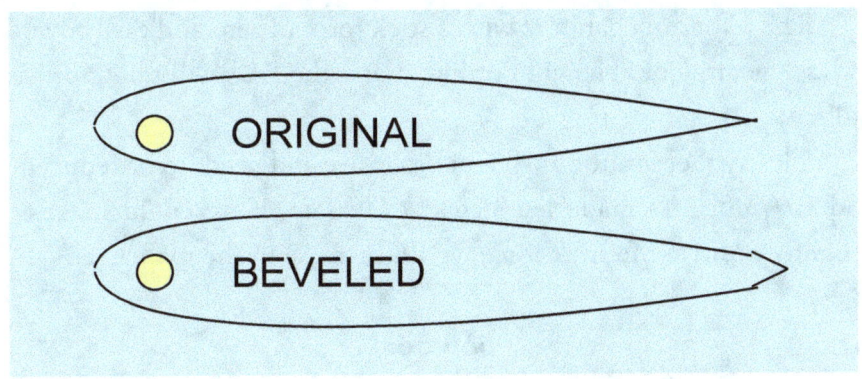

Figure 7.5

7.5.3 A SIMPLE DESIGN PROBLEM AND SOLUTION

Problem

Suppose an airplane has been designed such that it only takes 1 degree of trailing edge up elevator to increase the angle of attack by 2 degrees. Suppose also that the elevator control surface has the following hinge moment coefficient values:

$$Ch_\delta = -0.70 \text{ and } Ch_\alpha = -0.35$$

$$\text{Now } Ch = Ch_\alpha * \alpha + Ch_\delta * \delta$$

And, when $\alpha = 2$ degrees with $\delta = -1$ degree, the hinge moment is:

$$Ch = (1/57.3) * [(-0.35) * (2) + (-0.70) * (-1.0)] = \text{zero!!!}$$

The pilot can change α with a stick force of zero and easily over-stress the airplane. He will not like it and the FAA will not approve it!!

Clearly the elevator up-float is excessive and needs to be reduced. So Ch_{α} must be made less negative. In fact, if we could make it slightly positive the stick force would increase in the maneuver.

Solution

Consulting the data files we find that either a horn balance one-eighth of the tail semi-span or a 35 degree beveled trailing edge could add a $\Delta\, Ch_{\alpha} = 0.4$ / *radian*. The horn balance adds a $\Delta\, Ch_{\delta}$ = 0.35 / *radian* and the beveled trailing edge adds a $\Delta\, Ch_{\delta} = 0.4$ / *radian*

So with either solution, Ch_{α} would change from -0.35 to +0.05.

For the horn balance, Ch_{δ} changes from -0.70 to -0.35 and for the beveled trailing edge, from -0.70 to -0.30. Both these values are well away from the -0.10 design limit mentioned in Chapter 6, (page 65).

Then with the horn balance, when $\alpha = 2$ degrees and $\delta = -1$ degree:

$$Ch = (1/57.3) * [(+0.05) * (2) + (-0.35) * (-1.0)] = 0.008$$

And with the beveled trailing edge:

$$Ch = (1/57.3) * [(+0.05) * (2) + (-0.30) * (-1.0)] = 0.007$$

Ch_{α} has a small positive value for both solutions that will cause the stick force to increase slightly during the pull-up maneuver.

The choice of which solution to implement may then depend on the company's design precedent, estimated production cost, ease of manufacture, etc.

7.5.4 TABS

A tab is a small surface that is attached on a hinge at the trailing edge of a control surface as shown in Figure 7.6 below

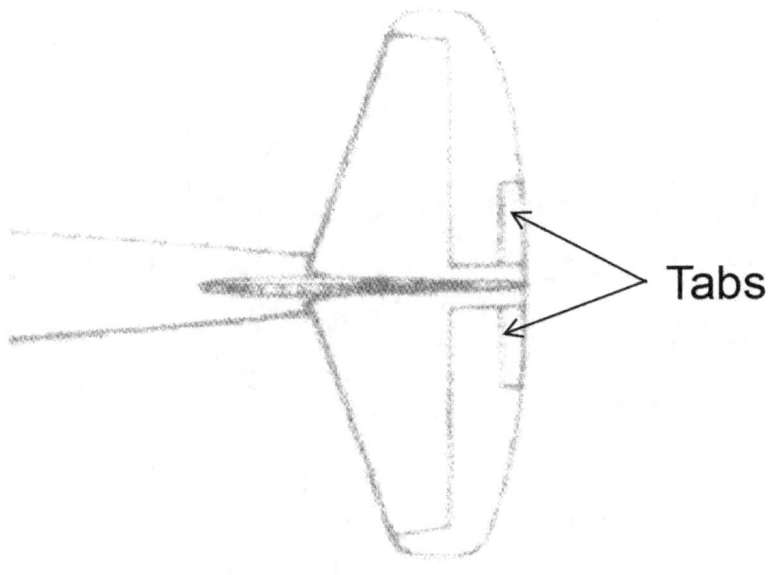

Figure 7.6

They may be used for a number of purposes on airplanes with un-powered control systems.

7.5.4.1 <u>TRIM TABS</u> ARE USED TO RELIEVE THE PILOT'S STICK FORCE BY HOLDING A CONTROL SURFACE IN ITS DEFLECTED POSITION. AN IRRE-VERSIBLE SCREW-JACK TRIM CONTROL IS USUALLY USED SO THAT THE TAB CANNOT FLOAT.

7.5.4.2 <u>BALANCING TABS</u> ARE USED TO MAKE THE CONTROLS FEEL LIGHTER. WHEN THE PILOT MOVES THE MAIN CONTROL SURFACE, A TAB CONNECTION SIMILAR TO THAT SHOWN IN FIGURE 7.7 MOVES THE TAB IN A DIRECTION THAT HELPS THE PILOT TO MOVE THE CONTROL SURFACE.

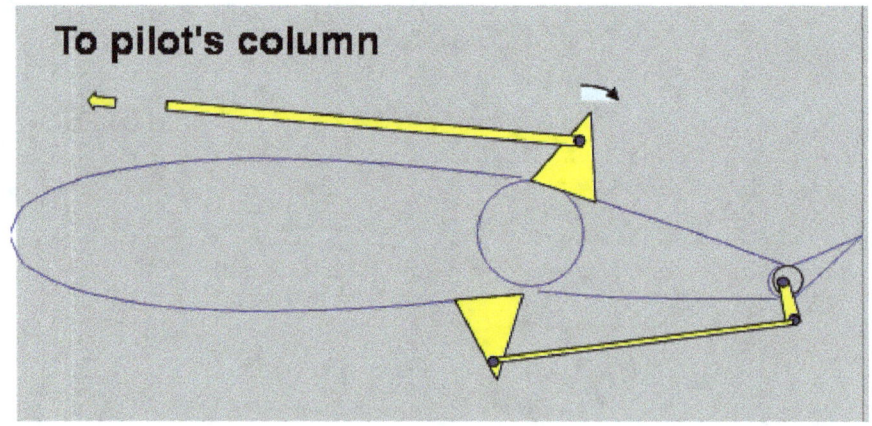

To pilot's column

Figure 7.7

7.5.4.3 <u>ANTI-BALANCE TABS</u> ARE USED TO MAKE THE CONTROLS FEEL HEAVIER. WHEN THE PILOT MOVES THE MAIN CONTROL SURFACE, A TAB CONNECTION SIMILAR TO THAT SHOWN IN FIGURE 7.8 MOVES THE

TAB IN A DIRECTION THAT RESISTS THE PILOT'S ATTEMPT TO MOVE THE CONTROL SURFACE.

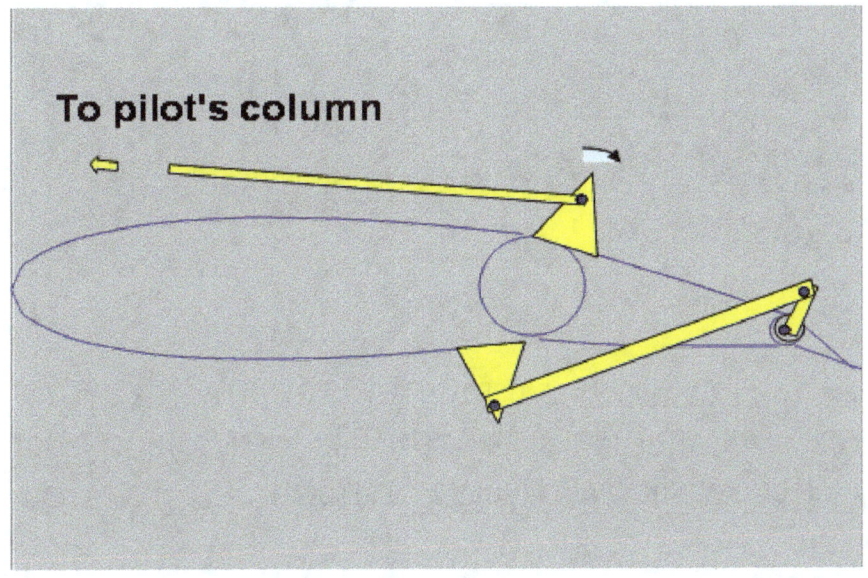

Figure 7.8

7.5.4.4 <u>CONTROL TABS</u>, SOMETIMES CALLED SERVO-TABS, ARE USED TO MOVE THE MAIN CONTROL SURFACE. THE MAIN CONTROL SURFACE IS FREELY HINGED AND THE PILOT'S CONTROL IS CONNECTED DIRECTLY TO THE TAB BY A LINKAGE THAT PASSES THROUGH THE MAIN SURFACES HINGE LINE AS SHOWN IN FIGURE 7.9

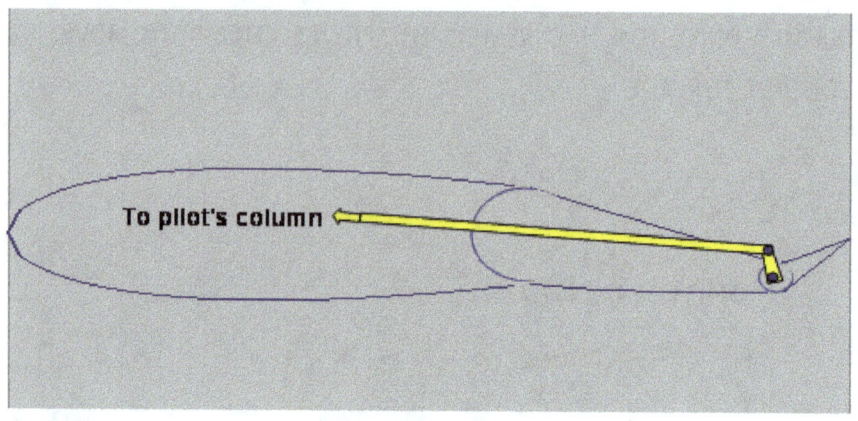

Figure 7.9

7.5.4.5 <u>SPRING TABS</u> ARE USED TO GET MORE HELP FROM THE TAB FOR MOVING THE CONTROL SURFACE AT HIGH AIRSPEED. THE CONTROL LINKAGE IS AS SHOWN IN FIGURE 7.10 BELOW:

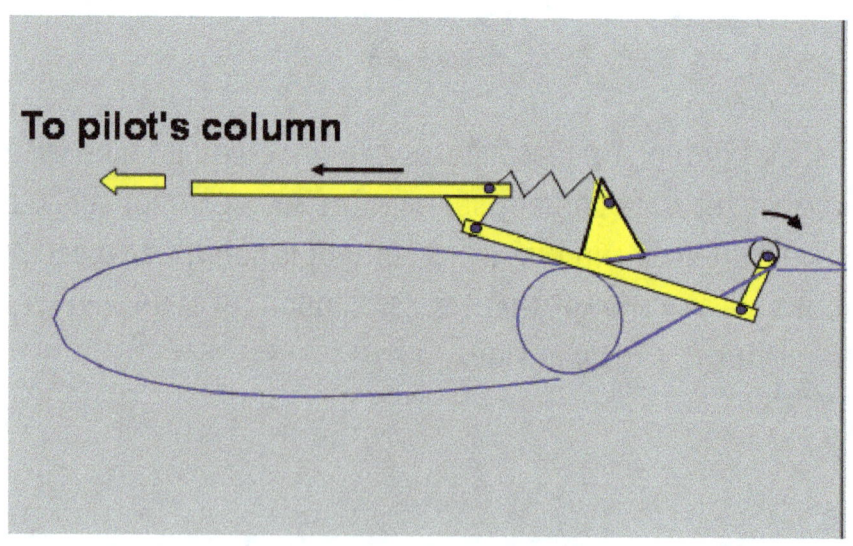

Figure 7.10

The pilot's control is connected to the control surface by a spring or torsion rod and to the tab by a control rod. At high airspeed the main control surface presents more resistance to being displaced and the consequent spring deflection causes more tab to be applied by the pilot's control.

Many types of tab control systems exist. The MD-80 airplane pitch control system uses three types of tab on the right and left elevators that extend over the full elevator span. The pilot's control is connected to a control tab on each half of the freely-hinged elevator. A balance tab on each half lightens the elevator hinge moment and a third tab prevents excessive elevator float when the adjustable tail is set at large leading-edge-down incidence angles. The un-powered tab control system results in acceptable pilot control forces on this medium-sized jet transport airplane. Table 6.2 shows that pilot's direct control of the elevator itself would require the elevator hinge moment to be about one third of the allowable -0.10 minimum value mentioned on page 65 of Chapter 6.

CHAPTER 8

Controls on Big Fast Airplanes

8.1 Hydraulic Power

The control surface hinge moments on big fast airplanes like the Douglas DC-10 and the Boeing B-747 are so large that we can no longer justify trying to avoid the use of hydraulic power in the control system design by using complicated methods to obtain the desired level of aerodynamic control surface balance. Instead, we use hydraulic pumps driven by the airplane's engines to generate hydraulic pressure in a number of hydraulic systems that drive actuators attached to the control surfaces. Then we connect the pilot's controls to a valve on each actuator that can be positioned to control the flow of hydraulic fluid into and out of the actuator. The pressures generated are typically 3,000 or 4,000 pounds/square inch and the hydraulic fluid flow rate is sufficient to move the control surface through its total travel in about one second. The control surfaces cannot float under the influence of aerodynamic pressures because the piston of the hydraulic actuator is effectively locked in its cylinder by the fluid that is trapped on each side of the piston. The controls are said to be "irreversible". The control stick cannot be moved by pushing up and down on the trailing edge of the elevator. Also, since the pilot has only to move a valve on the cylinder he cannot feel the aerodynamic hinge moment on the elevator. The force needed to move the valve is very small so that a synthetic force-feel system must be incorporated in the control system between the pilot's stick and the valve to provide stick forces that simulate the effect of airplane speed and control displacement.

8.2 ARRANGEMENT OF THE CONTROL ELEMENTS

An arrangement of the control elements and the basic principles of operation are explained below:

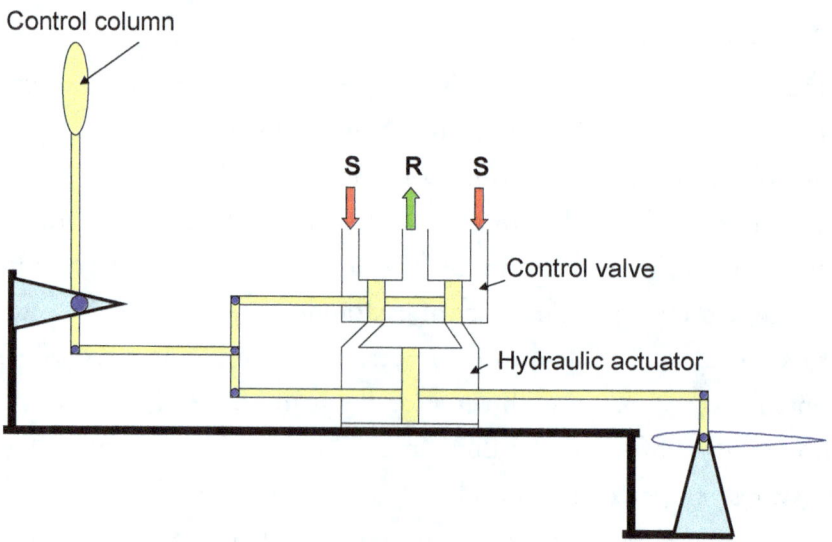

Figure 8.1

The control column is connected between a control valve and a hydraulic actuator as shown in figure 8.1 above. Pistons in the valve are aligned with access ports to the actuator and prevent hydraulic fluid flow from the high pressure supply, S and to the fluid return, R from the actuator. Both sides of the piston in the actuator are full of hydraulic fluid which is incompressible. So the piston cannot move unless the actuator ports are opened to allow fluid flow to and from the actuator's cylinder.

As shown in figure 8.2 below, movement of the control column causes the connecting bar between the control valve and the actuator to pivot about the actuator rod end pulling the lightly loaded control valve to the left. This allows a fluid supply at high pressure to enter the right hand cavity of the actuator cylinder and fluid from the left hand cavity to flow to return:

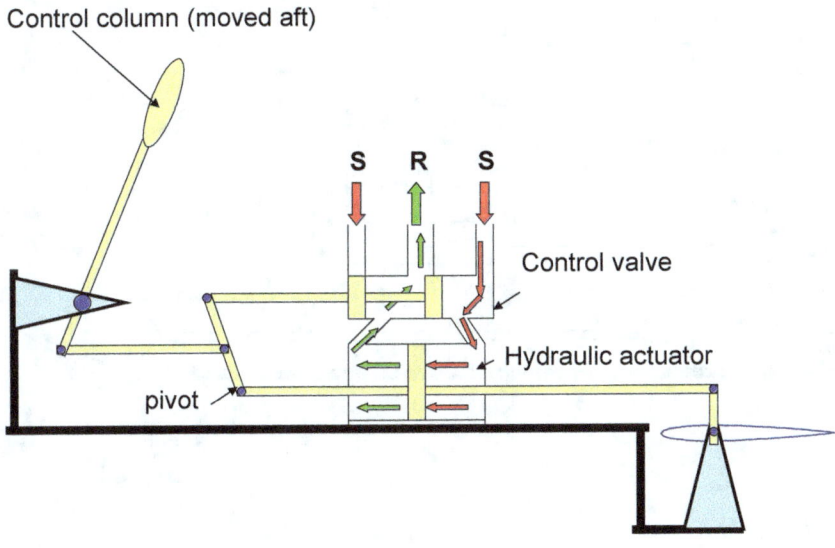

Figure 8.2

Then, with the control column held fixed the center of the interconnect bar acts as the pivot and as the high pressure on the actuator piston moves it to the left, the control valve pistons move to the right, closing the actuator access ports with the control surface deflected as commanded by the control column:

This is illustrated in figure 8.3 below:

The sequence of events described and illustrated separately in figures 8.1 through 8.3 occur smoothly and continuously so that the control surface movement always corresponds to the control column commands.

Piston cross sectional area is chosen so that sufficient force to move the control surface against its aerodynamic hinge moment can be produced from the available hydraulic pressure.

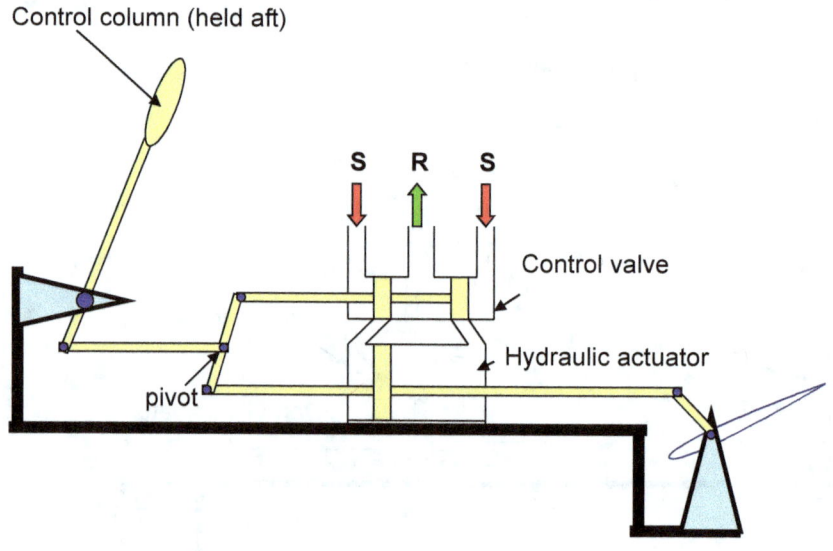

Figure 8.3

8.3 Designing for Safety

The probability of losing all hydraulic power to the control surface must be maintained at an acceptably low level. This is accomplished by careful assignment of multiple independent hydraulic systems to separate partial control surfaces in each control axis.

A typical arrangement is shown in figure 8.4.

Here the elevator is in four separate surface segments with each segment having its own hydraulic actuator and hydraulic pressure supplies.

Each of the airplane's three engines drive a hydraulic pump providing three independent sources of hydraulic power, yellow, blue and red, to the four elevator segments as shown.

If any single engine or hydraulic pump fails, then the elevator is still fully powered by the remaining two. Also, if any two engines or hydraulic pumps fail, at least two of the four elevator segments remain operable.

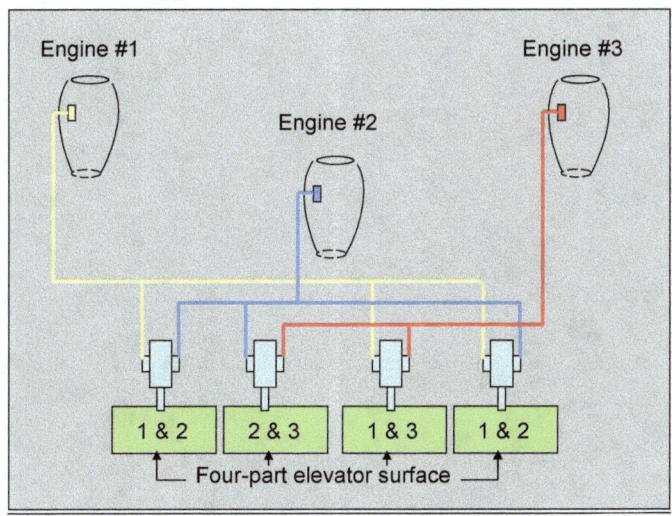

Figure 8.4

The operational status of the engines and the hydraulic systems is annunciated to the pilot who is trained to operate the airplane to maintain safe flight conditions with failed system elements.

CHAPTER 9

The Longitudinal Modes of Motion.

9.1 Rocking on a Pivot

In Chapter 4, the conditions for inherent longitudinal stability were described. But nothing was said about the type of motion that occurs when an inherently stable airplane recovers from a disturbance. So, we can say that Chapter 4 is simply an introduction to **static** longitudinal stability and in Chapter 9 we will begin to discuss **dynamic** longitudinal stability.

A simple way to start is to imagine a model airplane suspended in a wind tunnel on a pivot that passes laterally through its center of gravity.

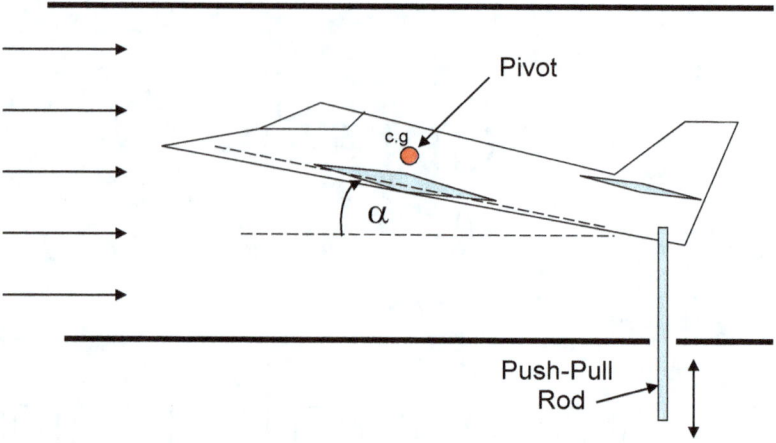

Figure 9.1
Airplane in wind tunnel pivoted through its c.g.

The model finds its equilibrium angle of attack, α, at which the aerodynamic moments about the pivot, (i.e. the c.g.), add up to zero.

A push-pull rod, attached at the rear, allows the test operator to rotate the model away from the angle of attack, α, at which it is in balance.

Because of the model's inherent static stability, the operator must pull to increase the model's angle of attack. Then if the rod is suddenly released from the model and removed from the tunnel, the model will move back towards its balanced angle of attack and may rock about the pivot with smaller and smaller movements until it settles in equilibrium. This oscillatory motion will involve angle of attack changes, oscillatory rates of pitch attitude and oscillatory accelerations in pitch attitude.

During the motion:

- The wing lift will oscillate due to the angle of attack changes. Due to its attachment to the pivot, the model cannot move vertically up and down as it would in free flight.
- The tail lift will vary due to:

  - The angle of attack changes,
  - The model's pitch rate,
  - The change of airflow downwash behind the wing that impacts the tail shortly after it leaves the wing trailing edge.

- The pitching inertia of the model will resist the pitch accelerations that occur.

The formulation of an equation to mathematically describe the oscillatory motion on the pivot must include all these effects and is a job for the professional aerodynamicist.

For now, we can simply say that three different types of moment are involved in the motion.

- M1 is the moment due to the model's inertial resistance to **rotational acceleration.**
- M2 is the moment due to the model's resistance to **rotational rate**.
- M3 is the moment due to the model's resistance to changing its **rotational angle** from its equilibrium angle.

Since the model is free to rotate on the pivot these three moments must find a balance during the motion so that, at all times during the oscillation,

$$M1 + M2 + M3 = 0$$

The ensuing mathematics produces a solution from which the dynamic stability of the "free-to-rotate" model on a pivot can be assessed.

The solution equation is shown graphically in three parts in Figure 9.2 below.

The first graph is simply a continuous rotational oscillation about the pivot which would occur if the model experienced no aerodynamic damping.

The second graph is a line that slowly decays from 1 to 0. This represents the decay of energy in the oscillation due to the actual presence of aerodynamic damping.

The third graph results from multiplying each value on the first by the value directly below it on the second. This produces the damped oscillatory motion with which the model settles back to its equilibrium angle of attack.

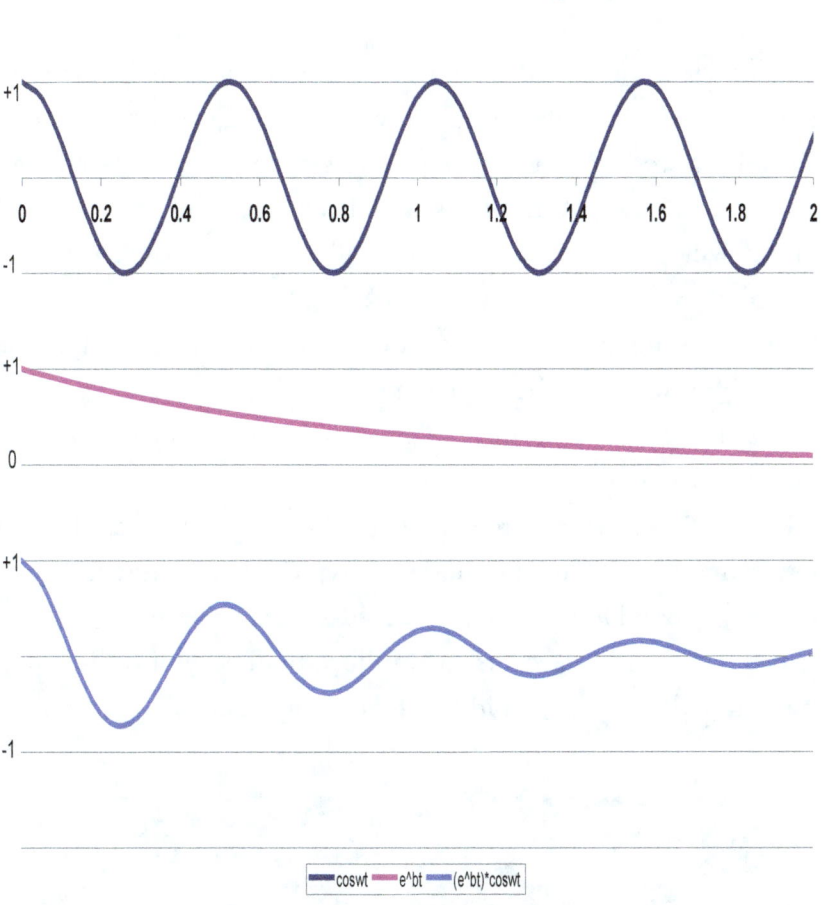

Figure 9.2

When the ordinate values of the lower curve are multiplied by the initial displacement, $d\alpha(0)$, we have the free motion time response, $d\alpha(t)$, as the model moves back to its equilibrium angle of attack.

9.2 AIRPLANE SHORT PERIOD MODE

Now, if instead of a model on a pivot, we visualize an airplane in flight, it too, will oscillate in pitch when its angle of attack is disturbed. However, since it has no pivot to constrain its motion, it will move up and down vertically as the lift changes and these vertical motions will also diminish with time like the pitch oscillations of the model.

This combined vertical and pitching motion is known as the **short period mode** of the airplane's longitudinal motion.

Airspeed changes during the short period are usually so small that they may be ignored.

Due to its larger size the full-scale airplane, it will oscillate at a lower frequency than the model. A typical response will appear as shown below: Here, in Figure 9.3 we see that the motion is well damped and rapidly recovers to equilibrium. The motion is sometimes referred to as the "rapid incidence adjustment".

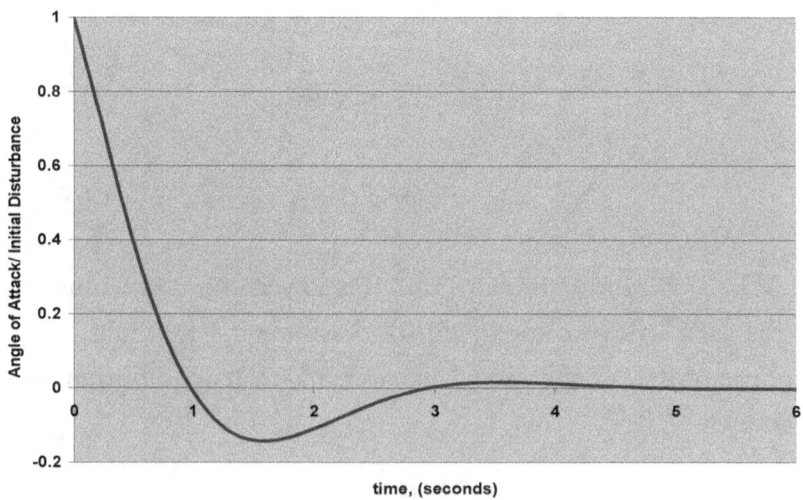

Figure 9.3

9.3 THE PHUGOID MODE

We ignored speed changes during the short period motion just described because the correction back to equilibrium angle of attack occurs so rapidly that there is little time for a speed change to occur. However, there is no constraint on the airplane's motion that will prevent the onset of a gradual speed change.

With no change of thrust or drag, the kinetic energy change due to the speed change must be compensated by a change of potential energy, (i.e. by a height change). So as speed increases height decreases. But the angle of attack is already stabilized at its equilibrium value; As speed builds up, lift increases, and the rate of descent is eventually arrested. The airplane then begins to climb and the speed begins to decrease.

Thus begins an oscillation known as the "phugoid", a name given to it by Frederick Lanchester, author of a book, "Aerodonetics", published in 1908; in it, he develops a theory of "pathways of airplanes". The theory is presented in a more recent text, "Theory of Flight" by Richard von Mises published in 1944 by Dover Publications.

The small thrust and drag changes that occur during the motion produce a damping influence so that the oscillation will decay very gradually.

The approximate time taken for one complete cycle of the phugoid oscillation is found to be:

$$t_P = \sqrt{2}\pi U_0 / g.$$

Where:

tP is the period in seconds and
Uo is the flight speed in ft/sec.

g is the gravitational acceleration = 32.2 ft/sec/sec.

At a flight speed of 400 ft/sec., tP = 55 seconds.

We see that the phugoid is a very slow oscillation.

Also, with controls fixed, the phugoid is slow to subside because a well-designed low-drag airplane cannot easily extract energy from the motion. It may be well to note however that, with the angle of attack remaining virtually constant, the airplane must pitch nose down to descend and nose up to ascend. Thus, the phugoid's speed and height oscillation also involves a pitch attitude oscillation that is easily prevented by the pilot who holds a constant pitch attitude when flying at a steady speed.

Since the phugoid is so slow and easily controlled by the pilot, it usually goes un-noticed. The short period oscillation however, must be rapidly suppressed to ease pilot workload and to prevent airplane damage.

The combined short period and phugoid free motion responses are in Figure 9.4 shown below:

We note the:

- Rapid subsidence of the angle of attack response.
- The ensuing low frequency, lightly-damped phugoid oscillation that involves changes of height and speed.
- The changing pitch attitude that is synchronized, (or in-phase), with the **slope** of the height trace.

The phugoid will be suppressed if the airplanes pitch attitude is held constant by the pilot or by an autopilot engaged in the "Pitch Attitude Hold" mode.

Note: *The student may check the equation for t_p given on page 92 by launching the model glider firmly, causing it to climb initially. Measure the distance covered, the total time taken, and count the number of oscillations that occur during its flight.*

The period, t_P, is the total time divided by the number of oscillations. The average speed, Uo, is the total distance traveled divided by the total time taken.

Use the equation to calculate the period for your measured value of Uo and compare the answer with the period you measured.

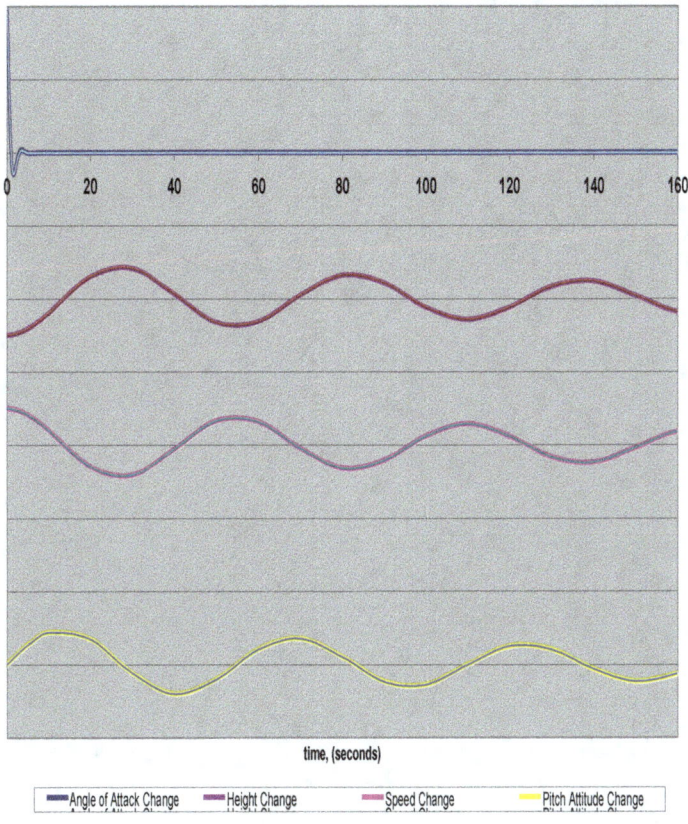

time, (seconds)

■Angle of Attack Change   ■Height Change   ■Speed Change   ■Pitch Attitude Change

Figure 9.4

CHAPTER 10

LATERAL-DIRECTIONAL AIRPLANE CHARACTERISTICS

10.1 THE DIFFICULTY

The reader will note from the chapter heading, that we do not introduce lateral and directional characteristics as two separate topics. This is because lateral motion, (rolling with one wing higher than the other), or sliding sideways, usually causes directional motion, (a change of heading). Likewise, a change of heading may cause a lateral skidding motion, that makes one wing rise above the other. Lateral and directional motions are inextricably connected and need to be studied as an entity.

During their early flights, the Wright brothers connected control wires to twist the wing tips of their glider asymmetrically for roll control. But when they applied wing twist to initiate turns, they found that the increased lift on the up-going wing, that was intended to cause a turn in the direction of the lateral tilt, actually increased the drag on that wing and caused the glider to turn away from the intended direction. So, the glider skidded in the direction of the down-going wing instead of turning towards it. A fixed vertical surface was attached a short distance behind the wings to resist the skidding motion and to turn the glider in the desired direction.

This design modification had the desired effect but, it degraded the ability of wing twisting to cause recovery from a banked turn. When the lower wing encountered the higher drag that was caused by the twist that was applied to lift it, it slowed, and the glider slipped towards the lower wing. Then, the fixed vertical fin caused the turn to tighten up.

As a result of these difficulties the Wright brothers decided to hinge the aft surface and to link its control to that used for wing twisting. The **lateral** control used for raising one wing above the other was linked to the **directional** control in such a way as to ensure the desired direction of heading change.

A photograph of Wilbur flying the glider with its aft vertical tail is shown in Figure 10.1 below.

Figure 10.1
(Ref: Google, "Wright Brothers' Glider Pictures")

10.2 ASYMMETRICAL FLIGHT

An airplane that is designed to carry a payload from one airfield to another spends most of its flight on a straight path with wings level. The airplane is in symmetrical flight with conditions on one side of its centerline plane being the same as those on the other. But events may occur during the flight that will require the airplane to adopt an asymmetrical orientation.

- At some point along its path it may be necessary to change direction.
- When arriving at the destination airport, there may be a wind blowing across the runway.
- Or a multi-engined airplane may experience an engine failure that causes a loss of thrust on one side, and leads to a sudden un-intended change of direction during flight or the take-off run along the runway.

In all these circumstances the airplane will be maneuvered into an **asymmetrical** orientation. In flight, it will fly with one wing tip higher than the other and possibly with some sideways skidding relative to the air.

10.2.1 TURNING FLIGHT.

To make a turn to the right, the pilot lowers the right wing and applies right rudder. Back pressure is applied to the control column to maintain altitude. If the correct amount of rudder is applied to erase any tendency for the airplane to skid sideways, then the turn is said to be coordinated. If the turn is not coordinated the ball in a "sideslip indicator", (shown in the lower right of figure 10.2), will be off-center. Here it indicates that a sideslip to the right exists. More right rudder is needed to coordinate the turn. The pilot learns to "step on the ball". He pushes on the right rudder pedal when the ball is off-center to the right.

In the coordinated turn at a bank angle of, φ, the lift, L, acts in the airplane plane of symmetry so that the vertical component of the lift, Lcosφ, supports the weight, W, and the horizontal component of the lift, Lsinφ, exactly balances the centripetal force

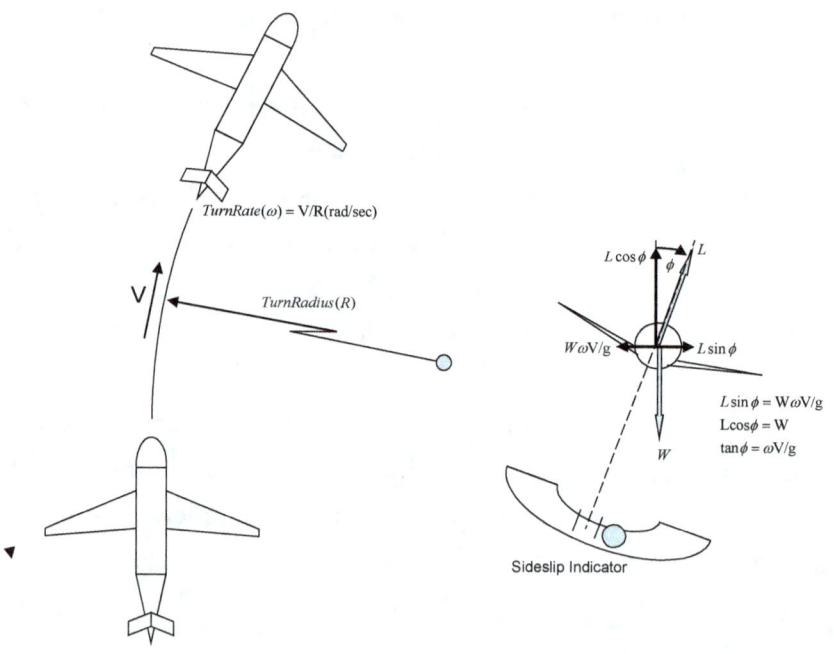

Figure 10.2

10.2.2 Crosswind Landing.

To prevent the crosswind from blowing the airplane off its course during an approach to a landing, the pilot banks into the wind as shown in Figure 10.3(a) below, or heads the airplane away from the runway centerline into the wind as shown in Figure 10.3(b). A combination of both these techniques may be used by some pilots.

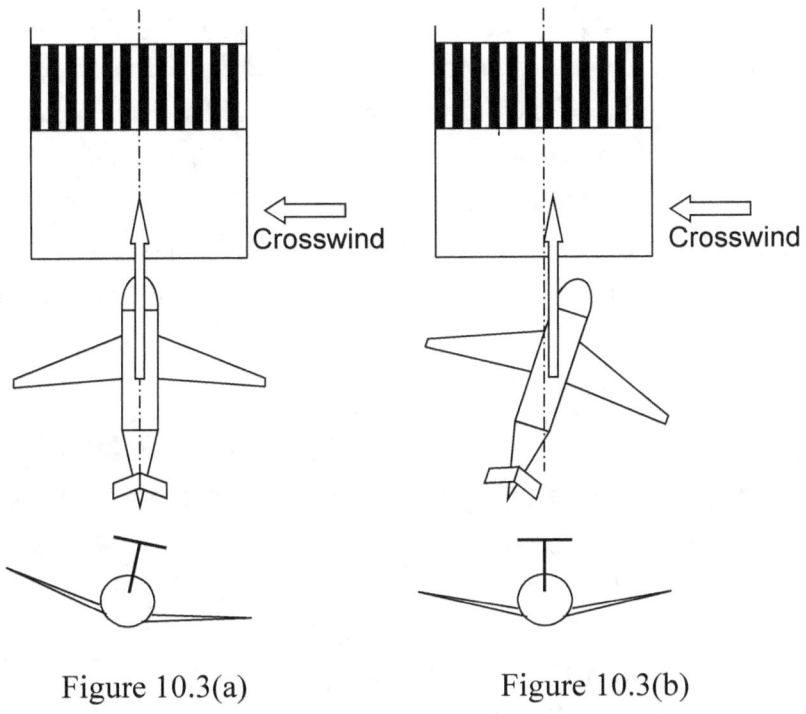

Figure 10.3(a)   Figure 10.3(b)

Figure 10.3

In most cases the aircraft will be controlled to fly upwind and parallel to the runway center line; so that when the bank angle or heading is corrected prior to touchdown, the crosswind will carry the airplane to land on or closer to the runway centerline.

10.2.3 ENGINE FAILURE.

When an engine that is offset from the airplane's centerline fails during take-off, full thrust must be maintained on the other engine(s) to attain the required climb gradient; the pilot must apply rudder to balance the yawing moment that is caused by the asymmetric thrust. When clear of the ground, the pilot will bank the airplane away from the inoperative engine so that the airplane's weight will counteract the side-force that is caused by the rudder deflection. If the bank angle must be limited, the airplane will side-slip until the total side-forces on the airplane are in balance. This is illustrated in Figure 10.4 below.

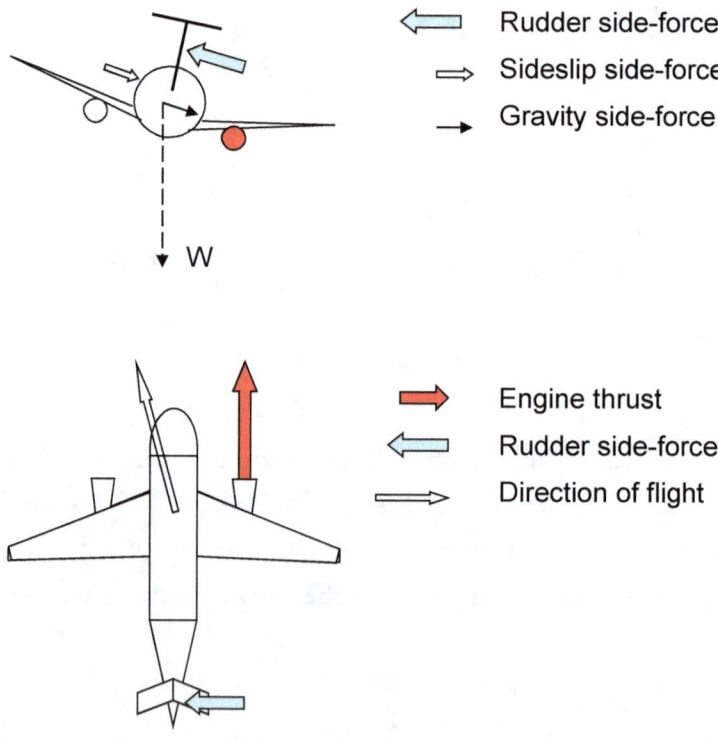

Figure 10.4

CHAPTER 11

Lateral-Directional Modes of Motion

11.1 Lateral-Directional Stability

Since we acknowledge that the airplane must sometimes be flown precisely in asymmetric maneuvers, we need to be assured that lateral-directional characteristics are such that the pilot's task to fly a desired path will not require extreme skill or cause undue fatigue. The lateral-directional characteristics of the airplane must be such that a minimum level of stability and controllability is maintained, and that control of asymmetric flight conditions that arise during crosswind landings and after engine failure should not be compromised by poor airplane flying qualities.

These and other considerations led to the formulation of quantitative flying qualities requirements initiated by pioneers Robert R. Gilruth in America and Sidney B. Gates in England.

Their initial research in the 1930's and '40's formed the foundation of formal requirements for several different classes of airplanes that appear to-day in Civil Airworthiness Requirements and in Military Standards.

11.2 LATERAL-DIRECTIONAL MOTIONS

Before we can evaluate lateral-directional stability, we need first to define the types of lateral-directional motion that can occur.

The freedoms that allow motion out of the plane of symmetry are:

a) Rolling about an axis, OX that is directed forward through the nose of the airplane.

b) Sliding sideways along an axis, OY that is directed to the right normal to the plane of symmetry.

c) Yawing about an axis, OZ that is directed downward through the floor of the airplane.

These axes originate at the airplane's center of gravity as shown in Figure 11.1 below:

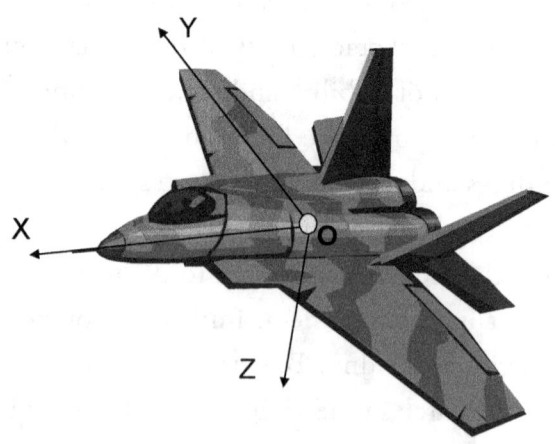

Figure 11.1

Motion velocity disturbance along the axis	OX	OY	OZ
Is given by the symbol	u	v	w
Angular displacement about the	ϕ	θ	Ψ
Rotational rate about the axis	p	q	r
Force along the axis	X	Y	Z
Moment about the axis	$\mathscr{L}$	$\mathscr{m}$	$\mathscr{n}$

Forces and velocity disturbances are taken as positive when they are in the direction of the axis. Angular displacements, rotational rates and moments are taken as positive when they are clockwise looking along the axis from its origin, O.

The motions in freedoms a), b), and c) above will not occur separately due to the interactive aerodynamic influence of one on the other. However, we can derive some useful descriptive information by imposing some restrictions, or limits, on motions to be studied as an initial step to our understanding of the completely free airplane.

11.3 Yawing on a Pivot

Perhaps the simplest motion to imagine is one in which the airplane is free to rotate in yaw about a fixed vertical pivot that passes through its center of gravity as shown in figure 11.2

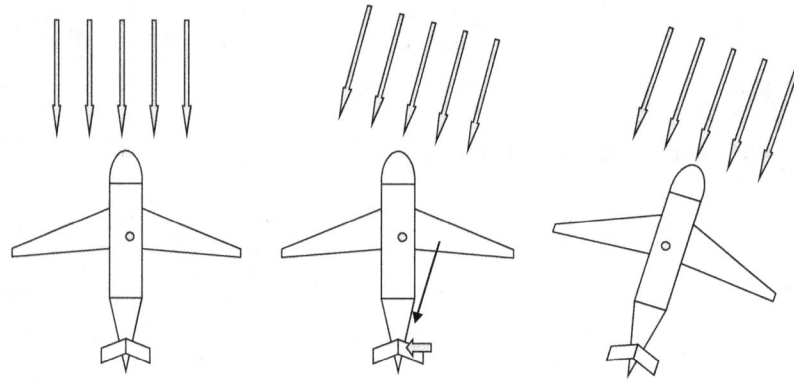

Figure 11.2

On the left, figure 11.2 a), the airplane is headed into wind. In the center, figure 11.2 b), the wind changes direction. The airplane is designed with an aft-mounted vertical fin that is sufficiently large to give it positive "weathercock stability". This causes it to rotate clockwise about OZ and to head into wind as shown in figure 11.2 c).

A positive sideslip, v, into the wind as shown in figure 11.2 b) has given rise to a positive yawing moment, N, so that the change of yawing moment, DN, due to the change of sideslip velocity, Dv, is itself positive. DN/Dv is positive for static weather cock stability.

11.4 ROLLING ABOUT A PIVOT.

We assume that the airplane is on a pivot aligned with the axis OX through its center of gravity. This prevents sideslip along OY and rotation about OZ when the airplane is banked through an angle, φ, as shown in Figure 11.3 below.

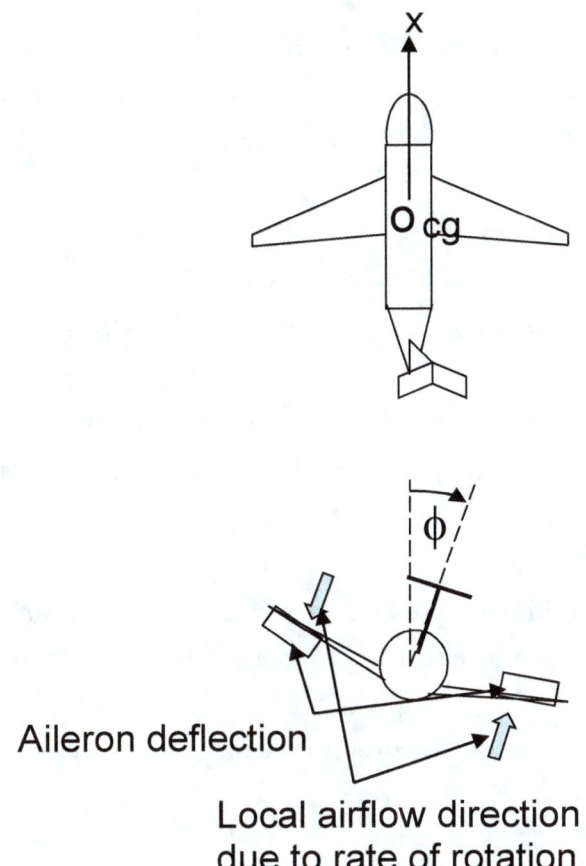

Aileron deflection

Local airflow direction
due to rate of rotation

Figure 11.3

The ailerons are deflected as shown through an angle, da causing the airplane to roll clockwise at a rate, p about the OX axis. The relative local airflow is as shown by the blue arrows causing the lift on the down-going right wing to increase and the lift on the up-going left wing to decrease. So, airplane roll rate, p causes a relative airflow that produces asymmetric lift that opposes the rotation rate being forced by the ailerons.

For equilibrium about a frictionless pivot the rolling moments must sum to zero. So the aileron moment that is forcing the motion must balance the rolling moment of inertia and the aerodynamic rolling resistance caused by the asymmetric lift.

The aileron moment is $\mathscr{L}(\delta a)$

And the aerodynamic rolling resistance is $\mathscr{L}(p)$

So in the final steady state, $\mathscr{L}(p) = \mathscr{L}(\delta a)$.

From known flight conditions and geometric properties of the airplane the effectiveness of the ailerons, $\Delta\mathscr{L}/\Delta\delta a$, can be determined so that $\mathscr{L}(\delta a) = (\Delta\mathscr{L}/\Delta\delta a) * \delta a$.

Also, the rolling resistance effectiveness, $\Delta\mathscr{L}/\Delta p$, (caused mainly by the wings), can be estimated so that $\mathscr{L}(p) = (\Delta\mathscr{L}/\Delta p) * p$.

Then the roll rate response to aileron deflection can be determined as follows:

$$(\Delta\mathscr{L}/\Delta p) * pss = (\Delta\mathscr{L}/\Delta\delta a) * \delta a$$

Giving $pss = \{(\Delta\mathscr{L}/\Delta\delta a)/(\Delta\mathscr{L}/\Delta p)\} * \delta a$

where 'pss' is the steady state roll rate reached due to continued application of the aileron deflection, 'δa'.

Following the sudden and sustained application of aileron deflection, the role rate, 'p' will build up to its steady state value 'pss' as shown in figure 11.4 below:

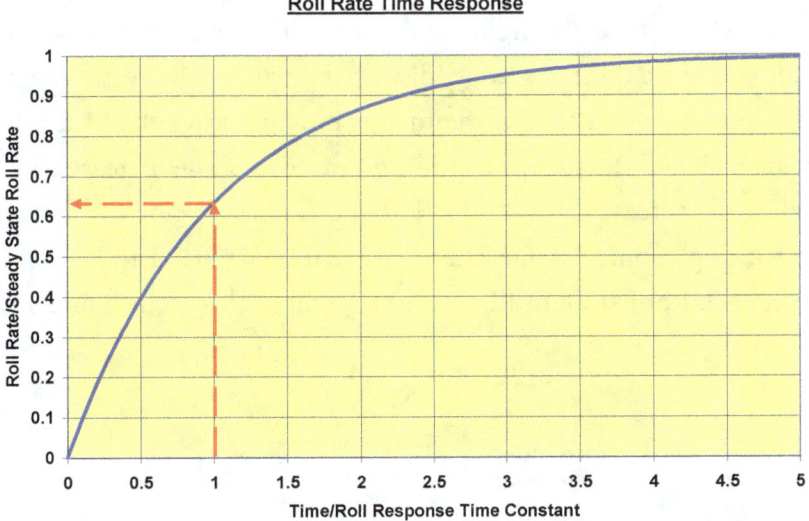

Figure 11.4

The rolling moment from the applied aileron deflection must overcome the mass moment of inertia of the airplane, Ixx in order to accelerate the roll rate to its steady state value. The delay is quantified by the Roll Response Time Constant, Ixx/(-ΔL /Dp).

Figure 11.4 shows that just over 60% of the steady state roll rate is reached in a time equal to the Roll Response Time Constant.

If the aileron input is suddenly returned to zero, then the roll rate will gradually slow to zero leaving the model airplane displaced in roll on the pivot at a steady state bank angle.

11.5 SLOW SPIRAL MOTION.

Now we should imagine that the yaw and roll pivots have been removed and the airplane is flying with no one at the controls. If the airplane banks slightly to the right, then it will begin to slip slowly to the right. Because of its positive weathercock stability, it will slowly turn to the right. If the side-slipping motion does not tend to correct the bank angle disturbance, it will increase because the wing on the outside of the turn is moving faster than the wing on the inside of the turn, and will gain more lift. The increased bank angle will increase the sideslip. The increased sideslip will increase the turning moment, and cause the resulting motion to be a slow divergence from straight flight into a spiral as shown in figure 11.5 below:

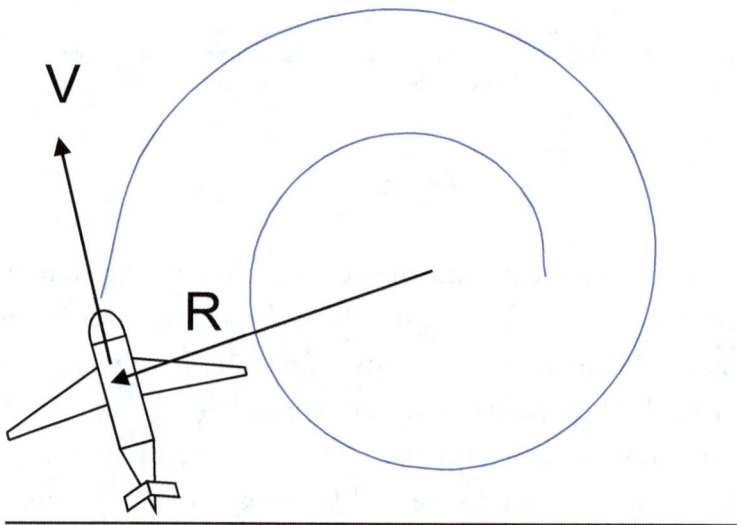

Figure 11.5

The aerodynamic influences that induce the spiral motion are:

a) The weathercock stability that produces a pro-turn yaw rate due to sideslip towards the center of the turn...... ($\Delta m / \Delta v$).

[A positive sideslip, v along the axis OY will cause a positive(clockwise) yawing moment, $\mathscr{N}$ about the axis OZ, so $(\Delta\mathscr{N}/\Delta v)$ will be positive.]

b) The rolling moment into the turn due to the asymmetrical lift caused by the yaw rate..................$(\Delta\mathscr{L}/\Delta r)$.

[A positive (clockwise) yaw rate, r about the axis OZ will cause a positive (clockwise) rolling moment, $\mathscr{L}$, about the axis OX so $(\Delta\mathscr{L}/\Delta r)$ will be positive.]

The influences a) and b) that support the spiral motion are both positive so their product $(\Delta\mathscr{N}/\Delta v) * (\Delta\mathscr{L}/\Delta r)$ must be positive.

The influences that resist the spiral motion are:

c) An aerodynamic yawing moment that opposes yaw rate.... $(\Delta\mathscr{N}/\Delta r)$

[A positive (clockwise) yaw rate, r about the axis OZ will be resisted by a negative (counter-clockwise) yawing moment, $\mathscr{N}$, about axis OZ so $(\Delta\mathscr{N}/\Delta r)$ will be negative.]

d) An aerodynamic rolling moment due to sideslip........... $(\Delta\mathscr{L}/\Delta v)$

[A positive sideslip, v along the axis OY will cause a negative(counter-clockwise) rolling moment, $\mathscr{L}$ about the axis OX, so $(\Delta\mathscr{L}/\Delta v)$ will be negative.]

The influences c) and d) that oppose the spiral motion are both negative so their product $(\Delta\mathcal{L}/\Delta v)*(\Delta\mathcal{M}/\Delta r)$ must be positive.

Now, it will be clear that, if the influences that support the spiral motion can overcome the influences that oppose it, then the motion will continue and will become more severe.

That is to say that the spiral mode of lateral-directional motion will be unstable if:

$$(\Delta\mathcal{M}/\Delta v)*(\Delta\mathcal{L}/\Delta r) > (\Delta\mathcal{L}/\Delta v)*(\Delta\mathcal{M}/\Delta r)$$

By conducting a more detailed mathematical analysis aerodynamicists can determine the time it takes for the bank angle and yaw rate to double in magnitude for an unstable spiral mode. The growth rate is usually slow and a slight instability is accepted.

A large vertical fin increases the yawing moment due to sideslip and contributes to spiral mode divergence while helping to dampen the dutch roll mode to be described next. Rolling moment due to sideslip acts to suppress the spiral mode but contributes to instability of the dutch roll mode.

11.6 THE DUTCH ROLL MODE.

This is an oscillation that involves all of the lateral-directional freedoms namely:

1. Rolling about the OX axis
2. Side-slipping along the OY axis and
3. Yawing about the OZ axis

These motions are strongly interdependent so that simplifications made to evaluate the mode by suppressing one of the freedoms gives only approximate technical results and presents an unrealistic view of the motion.

The most realistic view can be obtained from airplane flight test recordings taken when control inputs are made specifically to excite the lateral-directional motions. A flight recording taken during tests on a large commercial transport airplane to evaluate the dutch roll is shown in Figure 11.6 below.

Markings on the figure show that all the lateral directional modes are excited by aileron and rudder inputs that leave the controls undisturbed after about six seconds.

We see that the dutch roll oscillation is superimposed on the spiral mode that involves a sideslip of about two degrees and a yaw rate of about one degree per second. The roll mode is also excited and roll angle reaches a steady state value of about ten degrees after the aileron and rudder are centered.

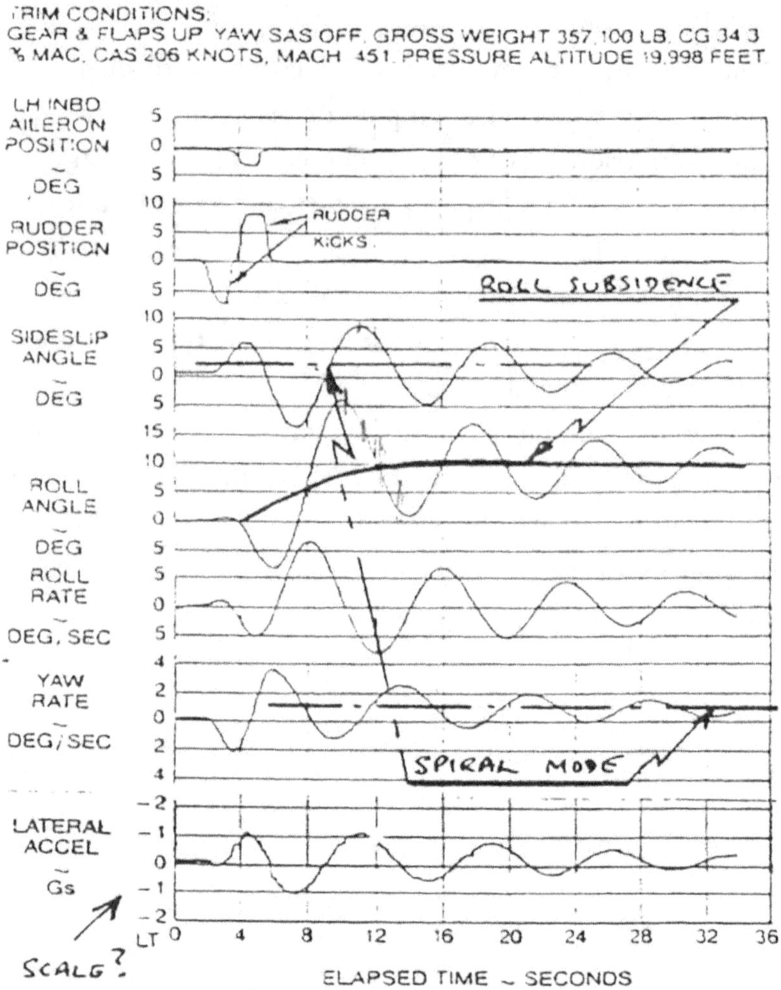

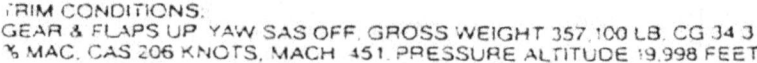

TRIM CONDITIONS:
GEAR & FLAPS UP YAW SAS OFF, GROSS WEIGHT 357,100 LB. CG 34.3 % MAC, CAS 206 KNOTS, MACH .451, PRESSURE ALTITUDE 19,998 FEET
LH INBD AILERON POSITION ~ DEG
RUDDER POSITION ~ DEG
RUDDER KICKS
ROLL SUBSIDENCE
SIDESLIP ANGLE ~ DEG
ROLL ANGLE ~ DEG
ROLL RATE ~ DEG/SEC
YAW RATE ~ DEG/SEC
SPIRAL MODE
LATERAL ACCEL ~ Gs
SCALE?
LT
ELAPSED TIME ~ SECONDS

Figure 11.6

In order to describe the dutch roll motion in more detail, the sideslip, yaw rate and roll angle traces have been redrawn and are presented in figures 11.7 thru 11.12 for the time period following the control excitation input. Figures 11.7 thru 11.9 describe the first half-cycle of the dutch roll free motion. The "controls-fixed" dutch roll oscillates about the spiral and roll subsidence motions sketched approximately in red.

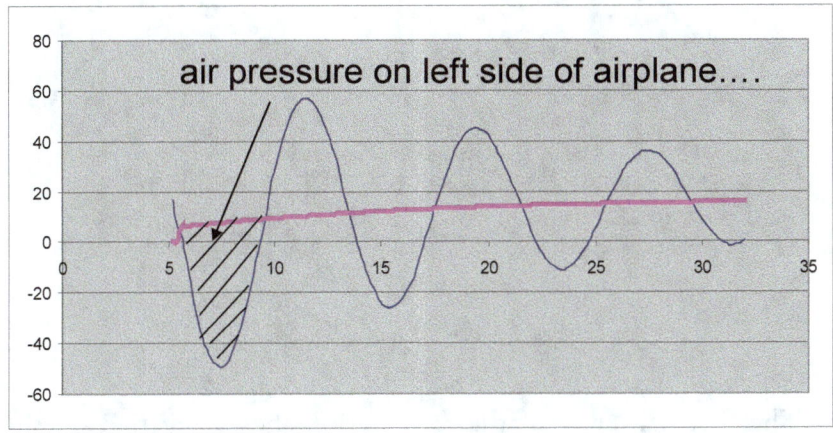

Figure 11.7- Sideslip Velocity, (ft/sec)

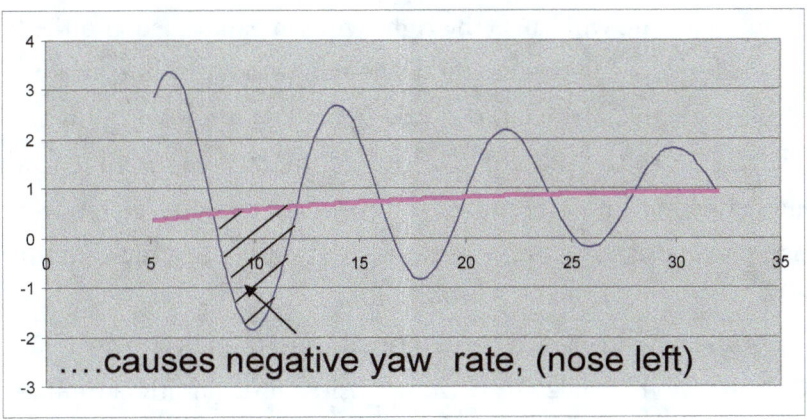

Figure 11.8 – Yaw Rate, (deg/sec)

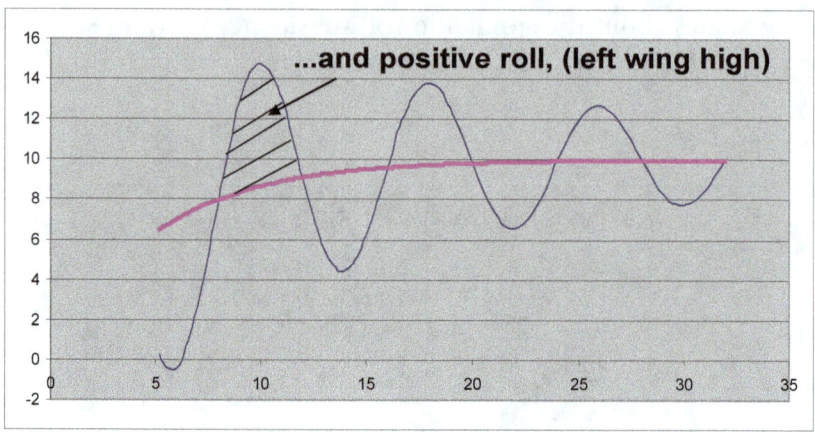

Figure 11.9 – Roll Angle, (deg)

The first half-cycle of the free motion shows negative sideslip velocity relative to the spiral motion sideslip. This causes a nose-left, (negative), yaw rate, (relative to the spiral mode yaw rate), due to the positive weathercock stability provided by the vertical fin. The nose left yaw rate will cause the right wing to gain speed and the left wing to lose speed, producing a left-wing-down rolling moment. However, on this airplane, the sideslip causes a left-wing-high roll angle relative to the roll angle of the roll mode response due to the wing's dihedral and sweepback. Too much dihedral effect would cause an excessive roll angle response and destabilize the dutch roll mode.

On the second half-cycle described in figures 11.10 thru 11.12 the sequence is repeated with opposite and slightly attenuated responses. Right, (positive) sideslip causes nose-right yaw rate and right-wing-high roll angle relative to the roll mode roll angle.

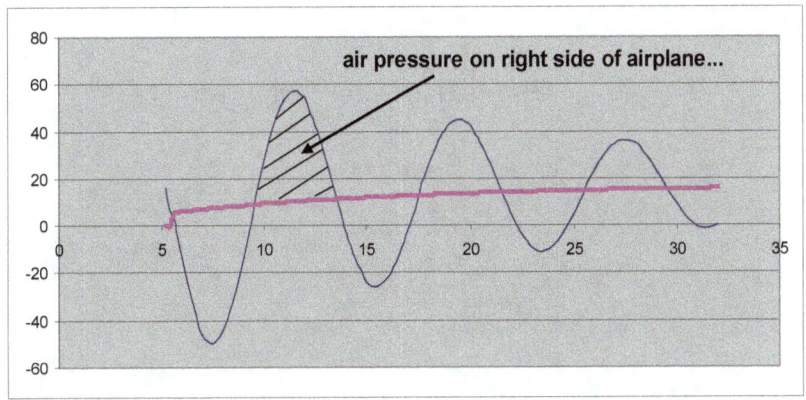

Figure 11.10- Sideslip Velocity, (ft/sec)

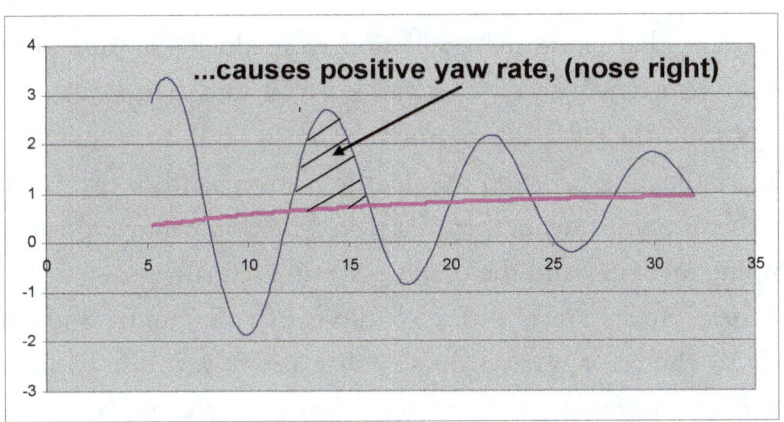

Figure 11.11 – Yaw Rate, (deg/sec)

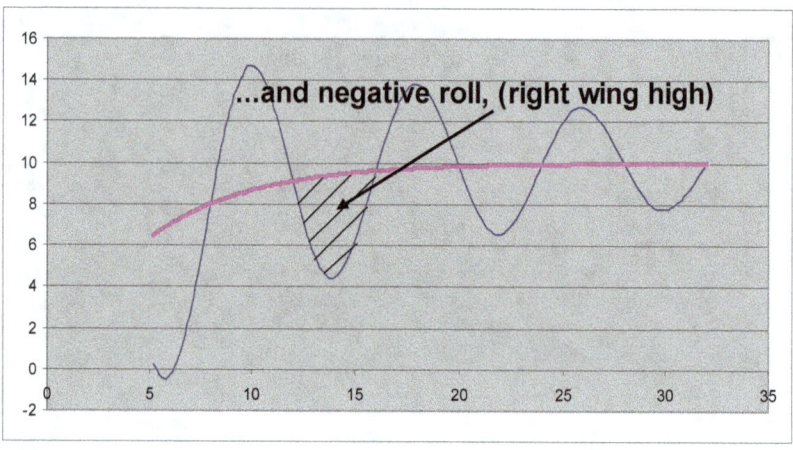

Figure 11.12 – Roll Angle, (deg)

Factors that cause the oscillation to gradually subside are the sideslip velocity, the yaw rate and the roll rate. These all extract energy from the cycle during the motion in both directions. If the rolling moment away from the sideslip is strong and/or the yaw rate into the sideslip is weak, then the forced motion will overpower the damping tendency and the oscillation will gradually grow.

These effects are opposite to those that destabilize the spiral mode so choice of vertical fin size that affects yaw due to sideslip and selection of wing dihedral and other geometrical features such as wing sweepback and wing height on the fuselage that affect roll due to sideslip is usually a compromise.

Figure 11.13 below depicts the opposing effects of airplane geometry on the spiral and dutch roll stability.

A slowly divergent spiral is normally acceptable and neutral stability of the dutch roll can be accepted. The dutch roll frequency is low with one cycle every five or eight seconds. Pilots are trained to control it. However, an automatic yaw damper system is usually implemented on swept wing airplane types that are more prone to have a low-damped dutch roll mode.

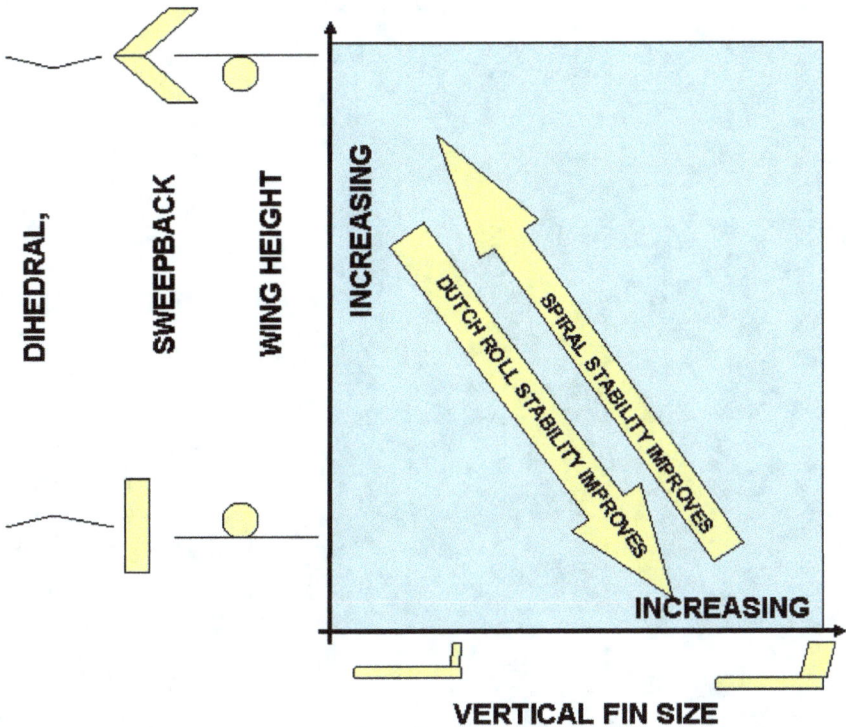

Figure 11.13

CHAPTER 12

Sizing the Tail

12.1 Basic Considerations.

The horizontal tail must serve two purposes:

1. It must produce sufficient aerodynamic force, 'tail lift', to **balance** all of the other aerodynamic and propulsive pitching moments acting on the airplane. This 'tail lift' may act upward or downward depending on whether a nose-up or nose-down moment is to be balanced. When balance is achieved the pitch axis of the airplane is said to be 'in trim'. The tail must produce the required lift without operating too close to its own stall condition. If ice protection is not provided, the 'tail lift' capability with ice accretions must be accounted for.

2. When angle of attack is disturbed from the trimmed flight condition, the tail must produce a change of tail lift that

is sufficient to produce a specified level of **static longitudinal stability.** The airplane will then be inherently stable and require less pilot effort to maintain control.

The tail sizes needed to produce balance and static stability are evaluated separately in the following sections.

12.2 BALANCE.

The airplane will be balanced in pitch when the moment about an axis parallel to OY is zero. Taking moments about the "wing-body" aerodynamic center as shown in figure 12.1 below produces a simple equation for the tail lift needed to balance the airplane in pitch.

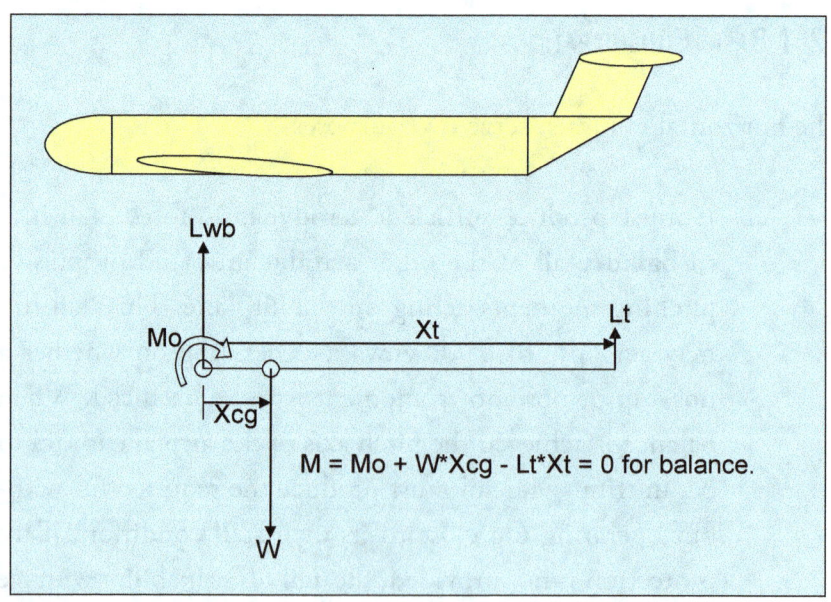

$$M = Mo + W*Xcg - Lt*Xt = 0 \text{ for balance.}$$

Figure 12.1

The 'in balance' equation can be re-written as follows;

$$Lt * Xt = Mo + W * Xcg$$

The tail lift, Lt, will be proportional to the tail area, St, so

$$Lt = K * St$$

and the balance equation can be written as

$$K * St * Xt = Mo + W * Xcg$$

Then the tail area required is

$$St = (Mo + W * Xcg) / (K * Xt)$$

$$= [(Mo / W) + Xcg] * [W / (K * Xt)]$$

From this we can see that for balance without a tail, (St= 0),

$$[(Mo / W) + Xcg] = 0$$

and

$$Xcg = -Mo / W$$

Then as the weight moves forward, a down load on the tail will be needed for balance. So 'K' will be negative and St increases by an amount equal to DSt = (W/K*Xt)*(ΔXcg)

A graph of the tail area, St, needed to balance the airplane with its center of gravity located at Xcg behind the "wing-body" aerodynamic center is shown in figure 12.2 below:

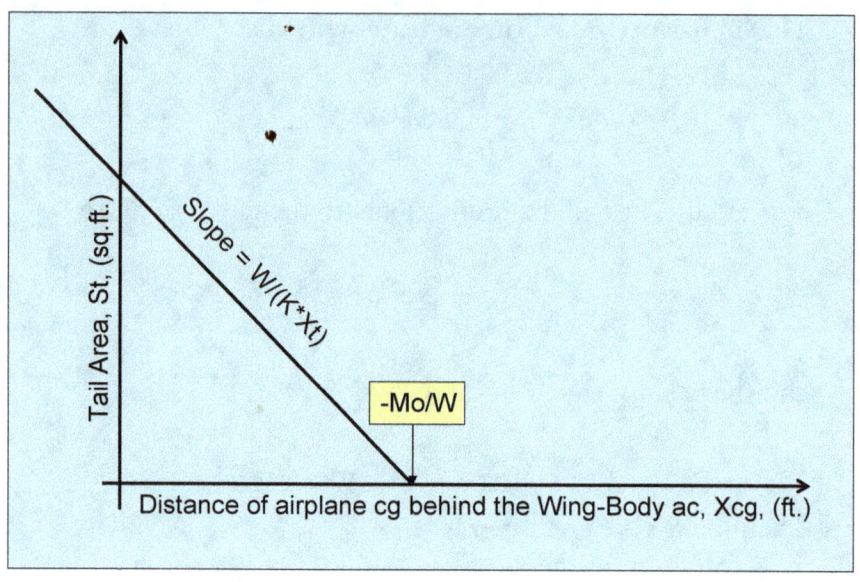

Figure 12.2

12.3 STATIC STABILITY

For static stability the change of pitching moment, ΔM, due to an increase of lift, ΔL, must be negative.

From figure 12.1 the pitching moment, M, about the wing-body aerodynamic center is:-

$$M = Mo + W*Xcg - Lt*Xt$$

and

$$W = Lwb + Lt = L$$

with

$$Lt = K*St$$

Then,

$$M = Mo + L*Xcg - K*St*Xt$$

And

$$\Delta M/\Delta L = Xcg - K'*St*Xt = 0 \text{ for neutral stability}$$

Note: $K' = \Delta K/\Delta L$ and represents the change of tail lift with total lift, L.

So, for neutral static stability the tail area required is:

$$St = Xcg/(K'*Xt).$$

The tail size required for neutral static stability can be graphed with the tail size for balance already graphed on figure 12.2 as shown in figure 12.3 below.

The cg location, Xcg must be to the right of the balance line in order for the tail to be capable of holding the nose up and on or to the left of the static stability line to prevent instability. So, for the tail to be capable of balancing the airplane at forward cg and stabilizing the airplane at aft cg, it must have an area, St that is above that at which the two lines cross over. The required tail size increases as the range between the forward and aft cg limits increase.

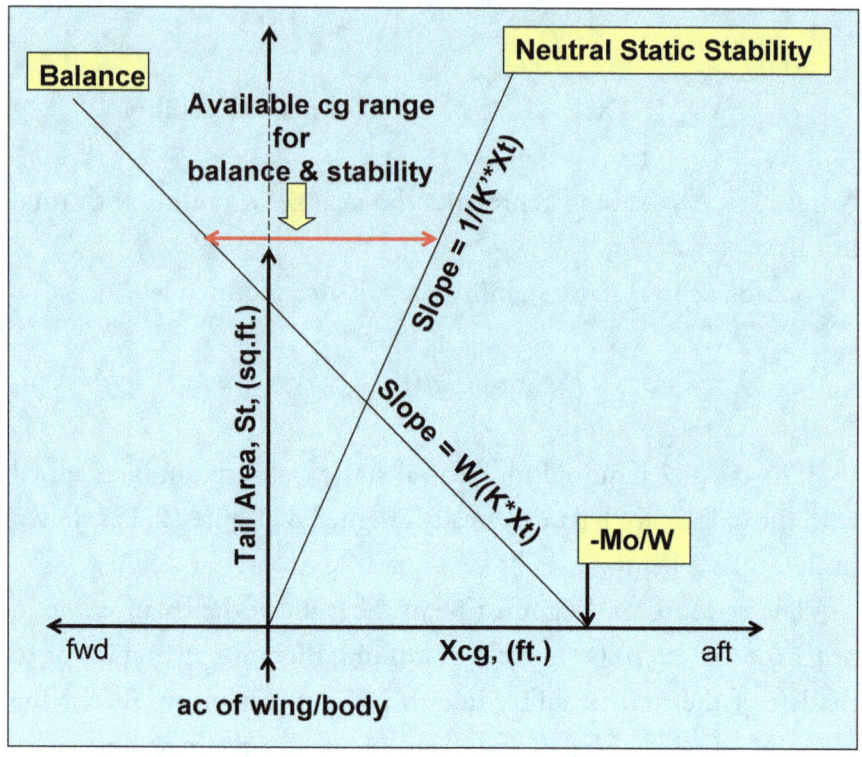

Figure 12.3

CHAPTER 13

Sizing the Canard

13.1 Basic Considerations.

The canard surface must produce sufficient aerodynamic force, 'canard lift', to **balance** all of the other aerodynamic and propulsive pitching moments acting on the airplane. This 'canard lift' will act upward to counteract a nose-down moment caused by the need to position the weight of the airplane well ahead of the wing-body aerodynamic center for inherent static stability

The canard size needed to produce balance and the cg location needed to ensure inherent static stability are evaluated in the following sections.

13.2 Balance.

The airplane will be balanced in pitch when the moment about an axis parallel to OY is zero. Taking moments about the "wing-body" aerodynamic center as shown in figure 13.1 below produces a simple equation for the canard lift needed to balance the airplane in pitch.;

$$M = Mo + W*Xcg + Lc*Xc = 0 \text{ for balance}$$

This balance equation can be re-written as follows:

$$Lc*Xc = - (Mo + W*Xcg)$$

The canard lift, Lc will be proportional to the canard area, Sc so;

$$Lc = K*Sc$$

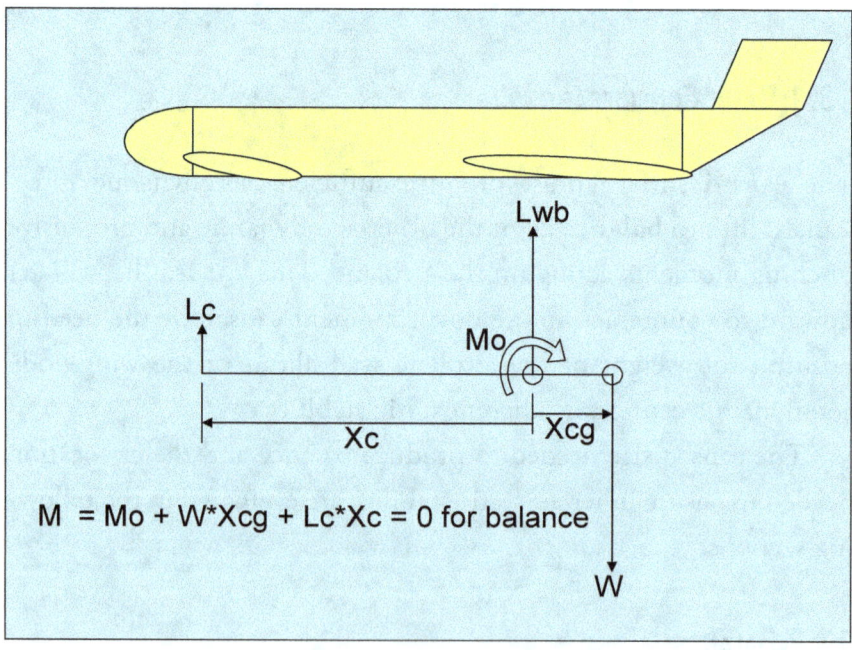

Figure 13.1

Then $\quad$ K*Sc*Xc = - (Mo + W*Xcg)

And the canard area required is

$$Sc = - (Mo + W*Xcg)/(K*Xc)$$

$$Sc = - [(Mo/W) + Xcg]*[W/(K*Xc)]$$

For balance without a canard, (Sc = 0), so

$$[(Mo/W) + Xcg] = 0 \text{ and}$$

$$Xcg = -Mo/W$$

Then as the weight moves forward, an up load on the canard will be needed for balance. So 'K' will be positive and Sc increases by an amount equal to

$$\Delta Sc = - [W/(K*Xc)]*(\Delta Xcg)$$

A graph of the canard area, Sc, needed to balance the airplane with its center of gravity located at Xcg behind the "wing-body" aerodynamic center is shown in figure 13.2 below:

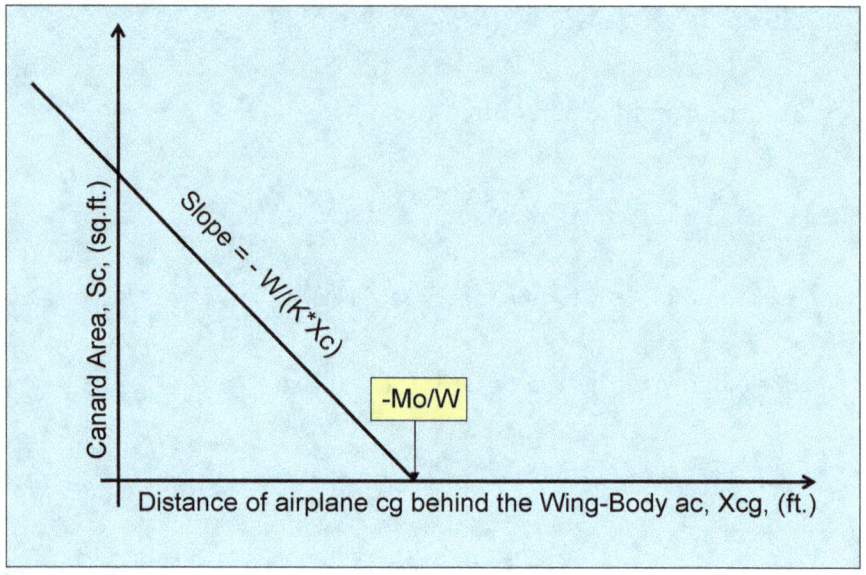

Figure 13.2

It will be noted that figure 13.2 is identical to figure 12.2 For providing balance, a down-loaded tail on moment arm Xt has the same effect as an uploaded canard surface on a moment arm Xc.

13.3 STATIC STABILITY.

From figure 13.1 the pitching moment, M about the wing-body aerodynamic center is:

$$M = Mo + W*Xcg + Lc*Xc$$

and $$W = Lwb + Lc = L$$

with $\qquad$ $Lc = K*Sc$

Then, $\qquad$ $M = Mo + L*Xcg + K*Sc*Xc$

and $\qquad$ $\Delta M/\Delta L = Xcg + K'*Sc*Xc = 0$ for neutral stability.

Where $K' = \Delta K/\Delta L$ and represents the change of canard lift with total lift, L.

So $\qquad$ $Sc = -Xcg/(K'*Xc)$ for neutral static stability.

The canard size, Sc required for neutral static stability is zero when Xcg is zero corresponding to the location of the wing-body aerodynamic center. Clearly, as Sc increases the cg must move forward to counteract the destabilizing effect of the canard surface. The equation can be graphed with the tail size for balance already graphed on figure 13.2 as shown in figure 13.3 below.

A comparison of tail and canard sizing results can be seen by combining the graphs on figure 12.3 with those on figure 13.3. This comparison is made in figure 13.4 and is discussed in the next section..

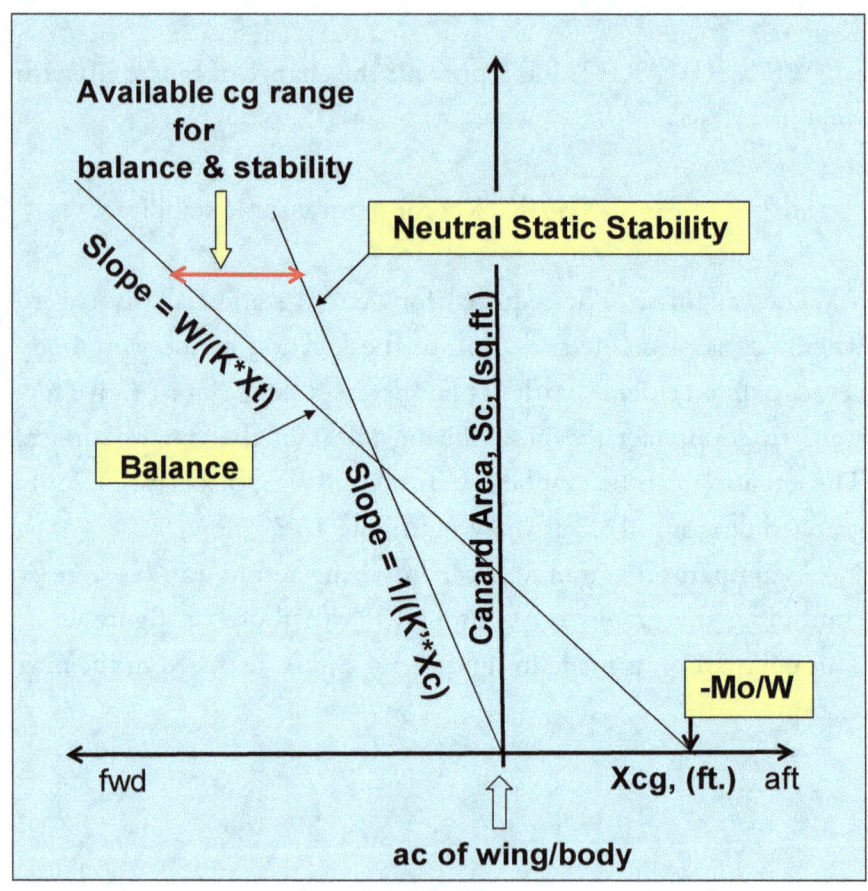

Figure 13.3

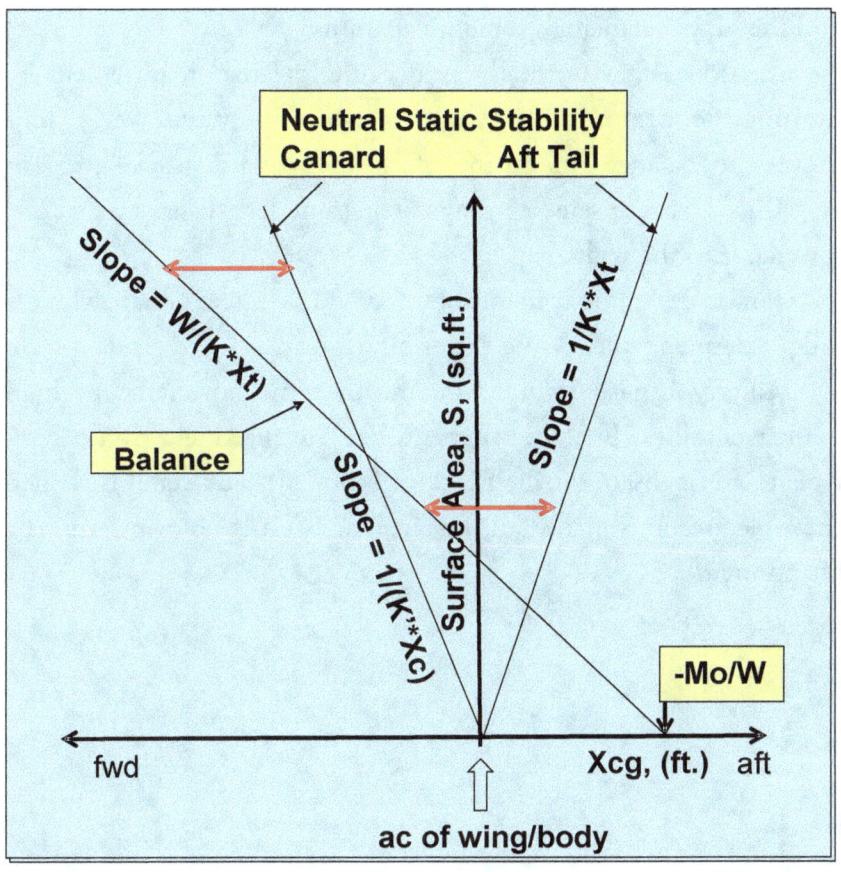

Figure 13.4

13.4 Canard/Aft Tail Comparison

Figure 13.4 shows that the inherently stable canard design must have a larger surface area to provide the same cg range, (red lines) and the same balancing moment capability.

Canard designs normally reduce the 'balance' requirement by limiting the maximum flap angle to reduce the value of Mo. This moves the 'balance' line to the left so that canard surface area can be reduced for the same cg range. The same length red line can be lowered into the wider gap.

Canard designs may incorporate some 15 degrees of flap, whereas tail-aft designs typically use about 40 degrees.

Military canard fighter aircraft have the advantage of high maneuverability, but they depend on advanced electronic controls to compensate for their lack of inherent static stability. When these electronic controls fail the airplane is un-flyable and must be abandoned.

CHAPTER 14

LOCATING THE WING ON THE FUSELAGE.

14.1 Basic Considerations

When deciding how far along the fuselage the wing should be located, the airplane designer will look for the position where changes in the amount of fuel and payload have minimal effect on the location of the center of gravity of the complete airplane. This will minimize the tail size required to balance and stabilize the airplane.

To demonstrate the effect of fore and aft wing location on an airplane's c.g. range, we will consider a commercial transport airplane that:

1. Has engines and landing gear attached to the wing
2. Carries all of its fuel in wing tanks,
3. Carries all passengers and payload in the fuselage, together with all airplane components that are not attached to the wing.

With a full passenger load and full fuel tanks, the airplane is at its Maximum Take-Off Gross Weight, (Max. TOGW).

14.2 COMMERCIAL CARGO TRANSPORT

Two missions will be considered:

1. A normal mission…..a long range flight with full payload.
2. A ferry mission…….a long range flight with zero payload.

In the first case the fuselage will be full and heavy and in the second case it will be empty and light. In each case the fuel tanks will be full when the flight begins and virtually empty if reserve fuel is burned before landing.

At the start of the normal mission, the airplane will have a full wing on a full fuselage and at the end it will have an empty wing on a full fuselage.

At the start of the ferry flight, it will have a full wing on an empty fuselage and at the end it will have an empty wing on an empty fuselage.

14.3 MASS PROPERTIES

Typical full and empty wing and fuselage weights and c.g.s are shown in figure 14.1.

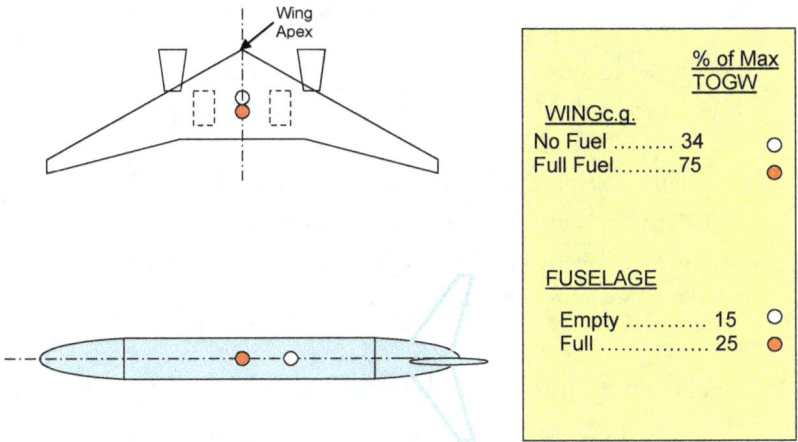

Figure 14.1

14.4 AIRPLANE CG VS. WING LOCATION

The airplane c.g. locations for all possible fore and aft wing locations are calculated using the equation shown in figure 14.2 The results are graphed in figure 14.3

Referring to Figure 14.2, the center of gravity location, $\bar{x}$, of the complete airplane with the wing apex at location, x, is found from:

$$W_{TOT} * \bar{x} = W_W * (x + \bar{x}_W) + W_F * \bar{x}_F$$

Giving

$$\bar{x} = \frac{W_W * (x + \bar{x}_W) + W_F * \bar{x}_F}{W_{TOT}}$$

Where:

$W_W = WingWeight$

$W_F = FuselageWeight$

$W_{TOT} = W_W + W_F$

$\bar{x}_W = WingCenterofGravityLocationbackfromWingApex$

$\bar{x}_F = FuselageCenterofGravityLocationbackfromFuselageNose$

$\bar{x} = TotalAirplaneCenterofGravityLocationbackfromFuselageNose$

$x = WingApexLocationbackfromFuselageNose$

We see from Figure 14.1 that the full wing represents about 75% of Max TOGW and we can expect that when it is moved along an empty fuselage the c.g. of the complete airplane will change significantly. Then if an empty wing is moved along a full fuselage the complete airplane c.g. movement will be less.

The results obtained for all four conditions are presented in Figure 14.3

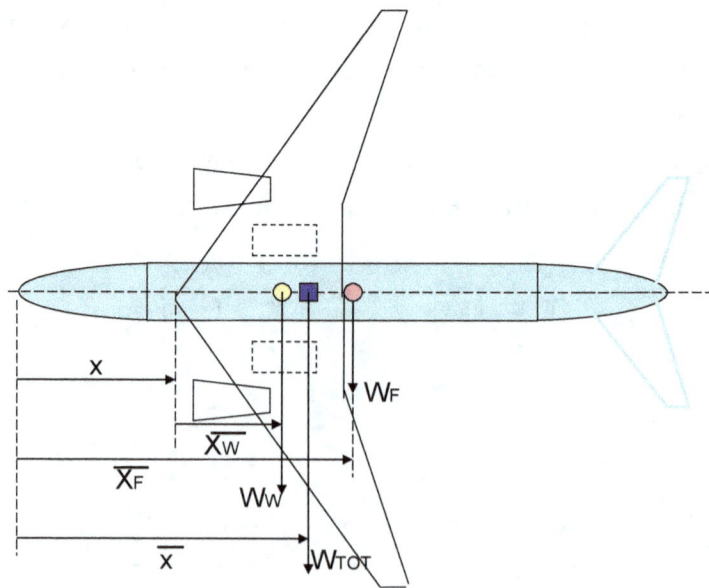

Figure 14.2

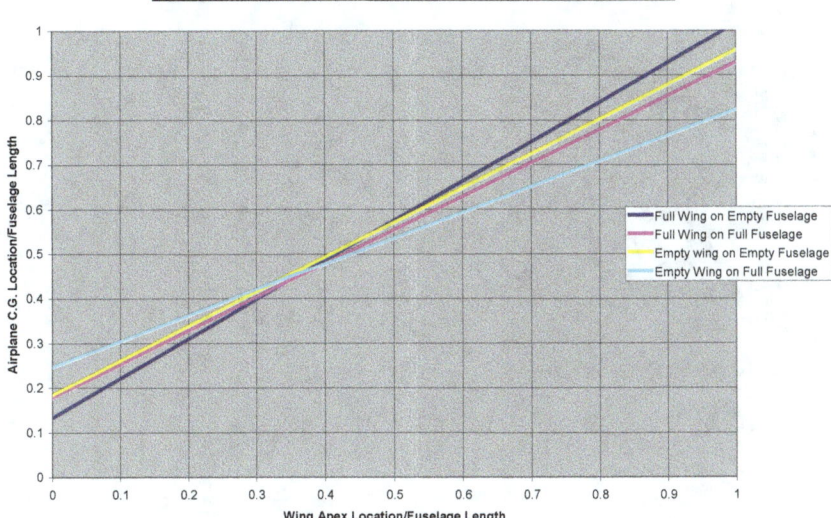

Effect of Wing Location on Total Airplane C.G. Location

Figure 14.3

14.5 Wing Location for Minimum CG Range

It is seen that the narrowest spread of c.g. locations that includes all four cases occurs when the wing apex is situated at about 35% of the fuselage length behind the airplane's nose.

A more precise conclusion can be derived from figure 14.4 that shows an expanded view of the narrow region.

This indicates that the narrowest c.g. range exists when the wing apex location is between about 0.34 to 0.40 of the fuselage length behind the airplane's nose, and that the c.g. range in that region is about 1.40% of the fuselage length. The fuselage length is 110 feet so the c.g. range required is110*0.014 = 1.54 ft..

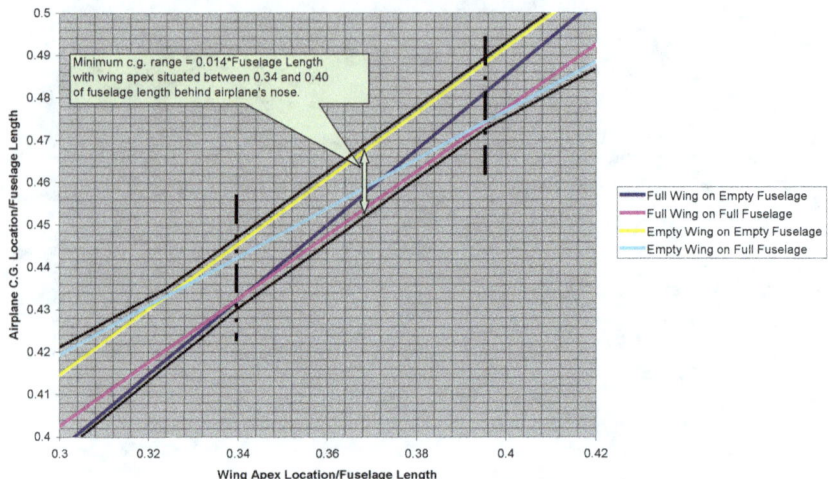

Figure 14.4

The next figure shows the wing apex set at 34% of the fuselage length back from the airplane nose giving forward and aft c.g. locations for the complete airplane of 43.2% and 44.6% of the fuselage length respectively.

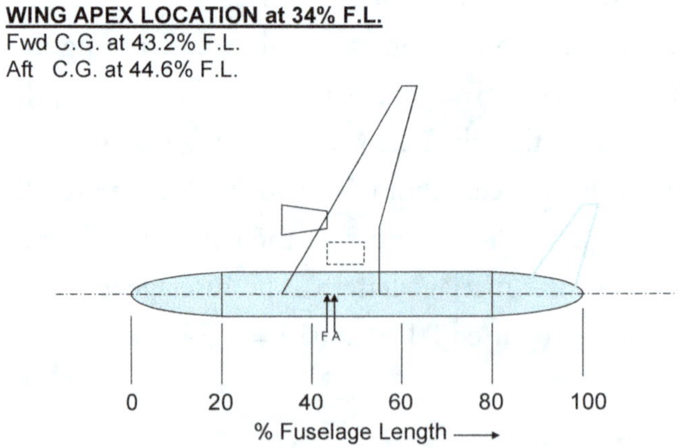

WING APEX LOCATION at 34% F.L.
Fwd C.G. at 43.2% F.L.
Aft   C.G. at 44.6% F.L.

Figure 14.5

REFERENCES

"Some Aeronautical Experiments" by Mr. Wilbur Wright, Dayton, Ohio, presented to The Western Society of Engineers, September 18, 1901.

"Boundary Layer Theory" by Dr. Hermann Schlichting Fourth Edition, published by McGraw Hill Book Company, 1960. (Results by Ludwig Prandtl)

"Control Surface Design in Theory and Practice" by M.B. Morgan, M.A., F.R.Ae.S., and H.H.B.M. Thomas, B.Sc., F.R.Ae.S., Journal of the Royal Aeronautical Society, No. 416, Vol. 49, August 1945.

"Airplane Stability and Control – A History of the Technologies That Made Aviation Possible" by Malcolm J. Abzug and E. Eugene Larrabee.

Cambridge University Press, 1997, ISBN 0 521 55236 2. (Chapter 5, Managing Control Forces)

ACKNOWLEDGEMENTS

The author wishes to acknowledge the benefits derived from the many comments and suggestions received from students during his twenty years of teaching at the California State University at Long Beach. Their open participation in the course work has helped to provide a focus for the book and to direct its content towards the very basic and frequently asked questions from young students studying to become professional aeronautical engineers.

Particular thanks are due to Ms. Chan Tran, a long-time co-worker on the C-17 program at Boeing, who relentlessly and meticulously, edited the text and identified areas where clarification was needed.